Strategies and Tactics for Management of Fertilized Hatchery Ponds

Strategies and Tactics for Management of Fertilized Hatchery Ponds

Richard O. Anderson
Douglas Tave
Editors

Food Products Press
An Imprint of
The Haworth Press, Inc.
New York · London · Norwood (Australia)

Published by

Food Products Press, 10 Alice Street, Binghamton, NY 13904-1580

Food Products Press is an Imprint of the Haworth Press, Inc., 10 Alice Street, Binghamton, NY 13904-1580 USA.

Strategies and Tactics for Management of Fertilized Hatchery Ponds has also been published as *Journal of Applied Aquaculture,* Volume 2, Numbers 3/4 1993.

Library of Congress Cataloging-in-Publication Data

Strategies and tactics for management of fertilized hatchery ponds / Richard O. Anderson, Douglas Tave, editors.

     p.   cm.

     Papers from a symposium held at the 1991 Annual Meeting of the American Fisheries Society in San Antonio, Tex.

     "Has also been published as Journal of applied aquaculture, volume 2, numbers 3/4 1993"–T.p. verso.

     Includes bibliographical references and index.

     ISBN 1-56022-048-1 (acid free paper).–ISBN 1-56022-049-X (pbk. : acid free paper)

     1. Fish ponds–Fertilization–Congresses. 2. Fish hatcheries–Management–Congresses. I. Anderson, Richard O. II. Tave, Douglas. III. American Fisheries Society. Meeting (1990 : San Antonio, Tex.)

SH159.S67 1994

639.3' 11–dc20

                          93-44926

                            CIP

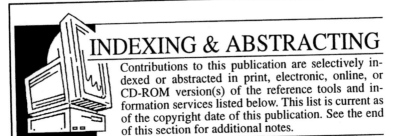

INDEXING & ABSTRACTING

Contributions to this publication are selectively indexed or abstracted in print, electronic, online, or CD-ROM version(s) of the reference tools and information services listed below. This list is current as of the copyright date of this publication. See the end of this section for additional notes.

(continued)

- *Environment Abstracts,* Bowker A&I Publishing, 121 Chanlon Road, New Providence, NJ 07974

- *Food Science and Technology Abstracts (FSTA),* scanned, abstracted and indexed by the International Food Information Service (IFIS) for inclusion in Food Science and Technology Abstracts (FSTA), International Food Information Service, Lane End House, Shinfield, Reading RG2 9BB, England

- *Foods Adlibra,* Foods Adlibra Publications, 9000 Plymouth Avenue North, Minneapolis, MN 55427

- *Freshwater and Aquaculture Contents Tables (FACT),* Food and Agriculture Organization, FIDI, (Att: MSCT/FACT), Via delle Terme di Caracalla, 00100 Rome, Italy

- *GEO Abstracts (GEO Abstracts/GEOBASE),* Elsevier/GEO Abstracts, Regency House/34 Duke Street, Norwich NR3 3AP, England

- *Marine Science Contents Tables (MSCT),* Food and Agriculture Organization, FIDI, (Att: MSCT/FACT), Via delle Terme di Caracalla, 00100 Rome, Italy

- *Referativnyi Zhurnal (Abstracts Journal of the Institute of Scientific Information of the Republic of Russia),* The Institute of Scientific Information, Baltijskaja ul., 14, Moscow A–219, Republic of Russia

- *Wildlife Review/Fisheries Review,* U.S. Fish and Wildlife Service, 1201 Oak Ridge Drive, Suite 200, Fort Collins, CO 80525–5589

(continued)

SPECIAL BIBLIOGRAPHIC NOTES

related to indexing, abstracting, and library access services

☐ indexing/abstracting services in this list will also cover material in the "separate" that is co-published simultaneously with Haworth's special thematic journal issue or DocuSerial. Indexing/abstracting usually covers material at the article/chapter level.

☐ monographic co-editions are intended for either non-subscribers or libraries which intend to purchase a second copy for their circulating collections.

☐ monographic co-editions are reported to all jobbers/wholesalers/approval plans. The source journal is listed as the "series" to assist the prevention of duplicate purchasing in the same manner utilized for books-in-series.

☐ to facilitate user/access services all indexing/abstracting services are encouraged to utilize the co-indexing entry note indicated at the bottom of the first page of each article/chapter/contribution.

☐ this is intended to assist a library user of any reference tool (whether print, electronic, online, or CD-ROM) to locate the monographic version if the library has purchased this version but not a subscription to the source journal.

☐ individual articles/chapters in any Haworth publication are also available through the Haworth Document Delivery Services (HDDS).

This book is for our wives,

Joan V. Anderson

and Katherine B. Tave

This book would have been impossible without their patience, understanding, support, and love.

Strategies and Tactics for Management of Fertilized Hatchery Ponds

CONTENTS

ABOUT THE EDITORS

Richard O. Anderson is retired from the U.S. Fish & Wildlife Service, National Fish Hatchery and Technology Center, where he was Senior Scientist. Dr. Anderson has been involved with fish hatchery research and management for many years. He was leader of the Cooperative Fishery Research Unit at the University of Missouri for over twenty years and previously served as Biologist in Charge at Wolf Lake State Fish Hatchery in Michigan.

Douglas Tave, of Urania Unlimited, has conducted innovative and seminal research in genetics, breeding, and reproduction of tilapia, catfish, baitfish, and shrimp. Dr. Tave is the author of the textbook *Genetics for Fish Hatchery Managers,* 2nd ed (Van Nostrand Reinhold) and the co-editor of the "Proceedings Auburn Symposium on Fisheries and Aquaculture." The author or co-author of over 80 professional and academic papers in aquaculture and related fields, Dr. Tave both wrote and directed the video *Aquaculture: Its Time Has Come* (Auburn Television, Auburn University). He is the founding editor of the Journal of Applied Aquaculture. He has been an aquaculture consultant to both the United States and Canadian governments.

Preface

This book contains a collection of papers that present new information on strategies and tactics that can be used to improve the ecological efficiencies of and economics of the production of fingerling fish, as well as shrimp, in fertilized ponds. Most of the papers were presented at a symposium on the management of fertilized hatchery ponds at the 1991 annual meeting of the American Fisheries Society in San Antonio, Texas.

Although intensive closed-loop aquaculture systems are receiving more and more attention, production of fingerlings in ponds will be the preferred and most effective system for most species. This efficiency is two-fold. First, ponds use solar energy to produce oxygen and to warm water; in addition, the natural microbial and algal communities in the pond recycle nutrients and waste products into fish food. In closed-loop systems, oxygenation and heating of water, and the feeding of larval diets or live food organisms are expensive. Equally important is the fact that fish that are produced for fishery management programs should be produced and cultured in systems that mimic, as closely as possible, the natural bodies of water into which they will be stocked. Fish that do well in closed-loop systems may be maladapted when stocked in lakes, rivers, and oceans.

In the future, the most effective hatchery managers will base management decisions on information that is site- and pond-specific, rather than blindly follow traditional generic recipes. Effective hatchery managers will develop management programs that will recognize how best to manage the dynamics of water quality in order to optimize fertilization rates and schedules in order to optimize the production of phytoplankton, zooplankton, and benthos; in addition, they will know how to avoid or to reduce problems caused by excessive photosynthesis, filamentous algae, blue-green algal blooms, rooted aquatic plants, and problem aquatic invertebrates.

xv

The papers in this book do not identify and solve all of the problems that confront hatchery managers. However, the papers do show that certain ecological problems have solutions, and they show the processes needed to identify and to solve others. It is hoped that this volume inspires others to expand upon the ideas that are explored so that hatchery managers in the future can anticipate and thus prevent problems rather than be forced to live with them or to react to them.

We especially want to thank Katherine B. Tave and Urania Unlimited for expert editorial assistance and for creation of the index.

Richard O. Anderson
Douglas Tave

New Approaches for Management of Fertilized Hatchery Ponds

Richard O. Anderson

ABSTRACT. This volume is a collection of papers that present some strategies and tactics that can be used to improve production and efficiency in the propagation of fingerlings in fertilized hatchery ponds. Species of fish in these papers include: paddlefish, *Polyodon spathula*; grass carp, *Ctenopharygoden idella*; silver carp, *Hypopthalmichthys molitrix*; bighead carp, *H. nobilis*; walleye, *Stizostedion vitreum*; saugeye, *S. canadense* ♀ × *S. vitreum* ♂; largemouth bass, *Micropterus salmoides*; striped bass, *Morone saxatilis*; hybrid striped bass, *M. saxatilis* ♀ × *M. chrysops* ♂. In addition, there is a paper discussing the culture of penaeid shrimp. It is proposed to designate the age of fish as: D1 = the day of hatch; D2 = the next day starting at midnight of D1; etc. It is also proposed to present fertilization amounts of nitrogen (N) and phosphorus (P) as μg/L or mg/L. Economic and ecological considerations lead to the recommendation that pond filling should be initiated as close to stocking as feasible and that the pond be filled after stocking. Water quality requirements of larvae vary with developmental stages and species. Excessive fertilization may cause elevated pH and toxic concentrations of un-ionized ammonia. Best results should be expected from fertilization on a pond-by-pond basis with plant meals or chemical fertilizers applied at appropriate concentrations of available N and P to achieve the desired responses.

Richard O. Anderson, National Fish Hatchery and Technology Center, United States Fish and Wildlife Service, 500 East McCarty Lane, San Marcos, TX 78666, USA. Correspondence may be addressed to 3618 Elms Court, Missouri City, TX 77459, USA.

[Haworth co-indexing entry note]: "New Approaches for Management of Fertilized Hatchery Ponds." Anderson, Richard O. Co-published simultaneously in the *Journal of Applied Aquaculture*, (The Haworth Press, Inc.) Vol. 2, No. 3/4, 1993, pp. 1-8; and: *Strategies and Tactics for Management of Fertilized Hatchery Ponds* (ed: Richard O. Anderson and Douglas Tave) The Haworth Press, Inc., 1993, pp. 1-8. Multiple copies of this article/chapter may be purchased from The Haworth Document Delivery Center [1-800-3-HAWORTH; 9:00 a.m. - 5:00 p.m. (EST)].

1

INTRODUCTION

Of all aquatic ecosystems, fertilized hatchery ponds are the most dynamic and challenging to manage. The standing crop of fish may go from less than 1 kg/ha at stocking to more than 100 kg/ha in about 5 weeks. In the early stage of larval development and growth, environmental requirements and limits of tolerance of fish may change rapidly. To achieve success, it is important to do the right thing at the right time. Management practices should sustain fertility and productivity, yet avoid problem growths of blue-green algae, filamentous algae, or rooted aquatic plants. The manager must achieve the production of the proper size, type, and amount of zooplankton and benthos to meet the needs of fish. To produce 100 kg of fish requires about 1,000 kg of fish food organisms. Since the normal rate of growth in weight of fingerlings is exponential, much more food is needed in the last half of the production period than in the first half. In the process of stimulating the production of food organisms, the manager must avoid excess densities of problem invertebrates that may function as larval predators or competitors of or for fish food organisms.

Management success and effectiveness are influenced by important decisions: (1) What kinds or sources of nutrients should be purchased? (2) When should pond filling be started, i.e., how much time and water are needed before fish are stocked? (3) What density and age of fish should be stocked? (4) How can a satisfactory quality of fish larvae and environmental variables be achieved so that larvae survive stocking and initiate normal feeding and growth? (5) Has the initial survival and growth been satisfactory, or should the pond be drawn down and restocked? (6) What kind and how much fertilizer should be added to a given pond today?

If a manager uses a systems approach to management, decisions are based on information. A key question is: What is the minimum amount of the most important information needed to make timely and proper decisions?

A systems approach is different from traditional fish culture where the manager has a recipe and every pond is treated in a similar manner. A weakness of this traditional approach is that ponds have individual "personalities." Stochastic and environmental variables are such that adjacent ponds are best managed as individual systems. The papers in this volume provide information about production studies that include several fish species: paddlefish, *Polydon spathula*; grass carp, *Ctenopharygodon idella*; silver carp, *Hypophthalmichthys molitrix*; bighead carp, *H. nobilis*; walleye, *Stizostedion vitreum*; saugeye, *S. canadense* ♀ × *S. vitreum* ♂; largemouth bass, *Micropterus salmoides*; striped bass, *Morone saxatilis*; hybrid striped bass, *M. saxatilis* ♀ × *M. chrysops* ♂. In addition, there is a paper on culture of penaeid shrimp. The results of these studies will help managers develop a systems approach to hatchery pond management. Major recommendations are summarized here for convenience.

SUMMARY OF RECOMMENDATIONS

What kinds or sources of nutrients should be purchased? Urea is recommended as the best source of nitrogen (N) (Opuszynski and Shireman 1993; Boyd and Daniels 1993; Anderson 1993a) and phosphoric acid as the best source of phosphorus (P) (Anderson 1993a); either alfalfa meal or rice bran is suggested as the best organic source of carbon, N, and P (Anderson 1993a; Mims et al. 1993; Summerfelt et al. 1993). An effective fertilization program may use chemical sources of N and P, or plant meal, or a combination. Chemical sources of nutrients may be much more cost effective than plant meals (Anderson 1993b; Culver et al. 1993).

How much time and water are needed before fish are stocked? Filling of ponds should be started as close to the time of stocking as feasible (Anderson 1993a; Culver et al. 1993; Opuszynski and Shireman 1993). Newly filled ponds are quickly invaded by problem invertebrates (Anderson 1993b; Burleigh et al. 1993; Czarnezki et al. 1993). There may be economic and ecological advantages to stocking ponds when they are only partially filled (Barkoh et al. 1993). Early filling may advance the time for development of problem growths of aquatic plants. However, ponds full over winter may avoid problems caused by excessive numbers of clam shrimp, *Cyzicus morsie* (Czarnezki et al. 1993).

What density and age of fish should be stocked? The best density of fish to stock depends on production targets, projected survival, and pond productivity for fish. By improving pond management,

Culver et al. (1993) increased the stocking density of walleye and saugeye larvae and increased fish production with little or no reduction in size of fish at harvest.

The best age of fish to stock varies with species and water temperature during development. The convention proposed for age of larvae is: D1 = the day of hatch; D2 starts at midnight of D1; etc. This is different from the convention where larvae hatched for 24-47 hours are considered to be 1 day old. A good age for stocking larval striped bass was D4, prior to the initiation of the critical period of inflation of the swim bladder (Anderson 1993b).

How can a satisfactory quality of fish larvae and environmental variables be achieved? The quality of larvae can be best assured if hatcheries have the facilities to hatch eggs and hold larvae at proper temperatures in water of good quality. Larvae should be well acclimated to the water quality of ponds. When larvae are stocked, pond temperature, dissolved oxygen, pH, and the concentration of un-ionized ammonia should be proper for the species and age stocked (Bergerhouse 1993; Opuszynski and Shireman 1993). The size, kind, and density of zooplankton present should be adequate for initiation of normal feeding and growth (Opuszynski and Shireman 1993). For striped bass, a crustacean density of 10-20/L at D5 was adequate (Anderson 1993b).

Has the initial survival and growth been satisfactory? A light at night and a small dip net were effective for sampling striped bass to evaluate initial survival and growth rate (Anderson 1993b). It is advantageous if a "go- or no-go" decision on whether to continue production or to start over can be made by D10. A light at night may also be used to make observations and subjective judgments on the status of zooplankton populations and on the relative abundance of some problem invertebrates.

A hatchery manager often has to decide on when to fertilize and on the kind and amount of fertilizer to add to a given pond. These decisions are critical for success and efficiency. Too much N or P can overstimulate photosynthesis and result in excessive pH and un-ionized ammonia during sensitive early larval stages (Anderson 1993a; Bergerhouse 1993). Inadequate application amounts and frequencies may result in deficient productivity. Objective concentrations of available N and P used by Culver et al. (1993) were

600 and 30 μg/L, respectively. Inorganic nutrients were added to ponds once a week to achieve these concentrations.

The convention adopted in these papers is that application amounts of N and P are expressed as mg/L or μg/L. Not all hatchery ponds have the same depth. Weekly application rates expressed as μg/l better describe the nutrients that are available for phytoplankton than do kg/ha or L/ha of fertilizer material or P_2O_5.

Published information on protein content was used to estimate available N in organic meals (% protein $\div$ 6.25 = % N); published information on P available for animal nutrition was used to estimate P (Table 1). Concentrations of P in chemical fertilizers were calculated on the basis that P = 43.7% of P_2O_5. Amounts of fertilizer needed to add given amounts of N or available P vary widely with the type of fertilizer (Table 2).

Lime should be used to increase buffering capacity of water with total alkalinity of <50 mg/L as $CaCO_3$. Important information on the amount of lime to add was provided by Boyd and Daniels (1993). They also documented the importance of urea as a source of N and of silicon to promote blooms of diatoms.

Managers should know the concentrations of nitrate nitrogen (NO_3^-–N), total ammonia nitrogen (TAN), and soluble reactive phosphorus (SRP) provided by their water supply and in ponds prior to fertilization and the stocking of larvae. They should know the density and taxa of zooplankton in their water supply and in ponds when larvae are stocked; this information helps determine whether to stock zooplankton. Dissolved oxygen and temperature data are needed in the morning and afternoon. Both variables can influence survival and growth of larvae. The change in oxygen from morning to mid-afternoon is an important estimate of net photosynthesis and plant response to fertilization (Anderson 1993b). Measurement of pH in the afternoon is important, especially in the early stages of development of sensitive species such as striped bass and walleye, because of the potential toxicity of elevated pH and un-ionized ammonia (Bergerhouse 1993). These measurements should be made daily, both prior to stocking and in the early stages of development. The concentration of TAN should be measured prior to fertilization to prevent excessive levels of ammonia. Knowledge of NO_3^-–N may also be important to calculate needed N. If SRP is

TABLE 1. Estimated concentration (%) of N and P in fertilizers applied to hatchery ponds. Values in section A are from Hubbell (1989); values in section B are from Piper et al. (1982).

Product	Nitrogen	Phosphorus Total	Available
A			
Alfalfa meal (20% protein)	3.2	0.27	0.22
Alfalfa meal (17% protein)	2.7	0.23	0.18
Alfalfa meal (15% protein)	2.4	0.22	0.17
Alfalfa meal (13% protein)	2.1	0.20	0.16
Brewers yeast	7.2	1.40	1.40
Corn	1.4	0.25	0.08
Cottonseed meal (50% protein)	8.0	1.20	0.50
Cottonseed meal[1] (41% protein)	6.6	0.90	0.30
Cottonseed meal[2] (41% protein)	6.6	0.95	0.32
Distillers solubles	4.5	1.35	1.20
Meat + bone meal (45% protein)	7.2	5.10	5.10
Meat + bone meal (50% protein)	8.0	4.10	4.10
Peanut meal[1]	6.9	0.55	0.20
Peanut meal[2]	7.2	0.60	0.20
Rice bran	2.0	1.50	0.23
Soybean meal[1]	6.7	0.60	0.20
Soybean meal[2]	7.0	0.60	0.20
Wheat shorts	2.6	0.40	0.13
B			
Alfalfa hay (15% protein)	2.4	0.24	-
Grass hay (6.9% protein)	1.1	0.21	-
Peanut hay (10% protein)	1.6	0.13	-
Chicken manure	1.3	0.40	-
Super phosphate	-	7.8-8.7	-
Triple super phosphate	-	19-22	-
Phosphoric acid	-	31.7	-
Ammonium nitrate	33.5	-	-
Anhydrous ammonia	82.0	-	-
Urea	46.0	-	-

[1]Expeller process
[2]Solvent process

expected to exceed $10\,\mu g/L$, analysis of this nutrient should also be measured prior to addition of phosphoric acid or other sources of P. Other information that may enhance management decisions includes: Secchi disk transparency; relative survival of larvae the week after stocking; swim bladder inflation; growth rate in mm/day to D8 or D10; size and relative abundance at D20 and D30; chlorophyll *a* concentration due to nanoplankton, i.e., the size of phytoplankton important to filter-feeding zooplankton; time and rate of development of filamentous algae and rooted aquatic plants.

TABLE 2. Ratios of N:P in selected plant meals and chemical fertilizers; application amounts needed to add 600 μg/L N or 30 μg/L available P. Pond mean depth for these calculations was 1.0 m.

Fertilizer	N:P	Nitrogen Percent in fertilizer	Nitrogen Amounts (kg/ha)	Phosphorus Percent in fertilizer	Phosphorus Amounts (kg/ha)
Plant meal					
Alfalfa meal	15:1	2.7	220	0.18	170
Cottonseed meal	22:1	6.6	91	0.30	100
Peanut meal	35:1	7.0	86	0.20	150
Rice Bran	9:1	2.0	297	0.23	130
Soybean meal	34:1	6.7	89	0.20	150
Chemical					
10-34-0	1:1.5	10.0	60	14.9	2.0
11-37-0	1:1.5	11.0	55	16.2	1.9
34-0-0	-	34.0	18	0.0	-
44-0-0	-	44.0	14	0.0	-
H$_3$PO$_4$ (75%)	-	0.0	-	31.7	0.9

The success of a systems approach by which management decisions are based on information will be two-fold: (1) an improvement in the quantity and quality of fish produced which will meet the needs of management programs and (2) an improvement in ecological and economic efficiencies.

REFERENCES

Anderson, R. O. 1993a. Apparent problems and potential solutions for production of fingerling striped bass, *Morone saxatilis*. Journal of Applied Aquaculture 2(3/4):101-118.

Anderson, R. O. 1993b. Effects of organic and chemical fertilizers and biological control of problem organisms on production of fingerling striped bass, *Morone saxatilis*. Journal of Applied Aquaculture 2(3/4):119-149.

Barkoh, A., R. O. Anderson, and C. F. Rabeni. 1993. Effects of pond volume manipulation on production of fingerling largemouth bass, *Micropterus salmoides*. Journal of Applied Aquaculture 2(3/4):151-170.

Bergerhouse, D. L. 1993. Lethal effects of elevated pH and ammonia on early life stages of hybrid striped bass. Journal of Applied Aquaculture 2(3/4):81-100.

Boyd, C. E., and H. V. Daniels. 1993. Liming and fertilization of brackishwater shrimp ponds. Journal of Applied Aquaculture 2(3/4):221-234.

Burleigh, J. G., R. W. Katayama, and N. Elkassabany. 1993. Impact of predation by backswimmers in golden shiner, *Notemigonus crysoleucas,* production ponds. Journal of Applied Aquaculture 2(3/4):243-256.

Culver, D. A., S. P. Maden, and J. Qin. 1993. Percid pond production techniques: Timing, enrichment, and stocking density manipulation. Journal of Applied Aquaculture 2(3/4):9-31.

Czarnezki, J. M., E. J. Hamilton, and B. A. Wagner. 1993. Water management to control clam shrimp, *Cyzicus morsie,* in walleye, *stizostedion vitreum,* production ponds. Journal of Applied Aquaculture 2(4):235-242.

Hubbell, C. H. 1989. Feedstuffs analysis tables. Feedstuffs, February 20:36.

Mims, S. D., J. A. Clark, J. C. Williams, and D. B. Rouse. 1993. Comparisons of two by-products and a prepared diet as organic fertilizers on growth and survival of larvae paddlefish, *Polyodon spathula,* in earthen ponds. Journal of Applied Aquaculture 2(3/4):171-187.

Opuszynski, K. K., and J. V. Shireman. 1993. Strategies and tactics for larval culture of commercially important carp. Journal of Applied Aquaculture 2(3/4):189-219.

Piper, R. G., I. B. McElwain, L. E. Orme, J. P. McCraren, L. G. Fowler, and J. R. Leonard. 1982. Fish Hatchery Management. United States Fish and Wildlife Service, Washington, D.C.

Summerfelt, R. C., C. P. Clouse, and L. M. Harding. 1993. Pond production of fingerling walleye, *Stizostedion vitreum,* in the northern Great Plains. Journal of Applied Aquaculture 2(3/4):33-58.

Percid Pond Production Techniques: Timing, Enrichment, and Stocking Density Manipulation

David A. Culver
Sharook P. Madon
Jianguang Qin

ABSTRACT. Tremendous variation in survival and growth of percid fry stocked in ponds was addressed through manipulation of amounts and kinds of fertilizers added and stocking densities of fish. Ponds were filled with water from nearby eutrophic lakes less than one week prior to stocking. Survival in these ponds averaged 64%, whereas ponds filled one month before stocking averaged only 14%. Optimal inorganic fertilization was identified as weekly restoration to 600 μg N/L (NH$_4^+$ + NO$_3^-$) and 30 μg P/L as PO$_4$. Organic fertilization gave variable results and water qualities. In one experiment, low organic fertilization alone (28 kg alfalfa meal/ha/week) provided survival and growth comparable to optimal fertilization with inorganic nutrients. Although similar in results, inorganic fertilization was more cost effective than the alfalfa meal. Low fish predation from low stocking or survival allowed *Daphnia* to overgraze algae. This resulted in a crash in both algae and zooplankton in the ponds. Doubling the initial stocking density increased harvests from 150,000 to 300,000 fish/ha and increased yield from 45 to 100 kg/ha.

David A. Culver, Sharook P. Madon, and Jianguang Qin, Department of Zoology and Ohio Cooperative Fisheries and Wildlife Research Unit, The Ohio State University, 1735 Neil Avenue, Columbus, OH 43210, USA.

[Haworth co-indexing entry note]: "Percid Pond Production Techniques: Timing, Enrichment, and Stocking Density Manipulation." Culver, David A., Sharook P. Madon, and Jianguang Qin. Co-published simultaneously in the *Journal of Applied Aquaculture*, (The Haworth Press, Inc.) Vol. 2, No. 3/4, 1993, pp. 9-31; and: *Strategies and Tactics for Management of Fertilized Hatchery Ponds* (ed: Richard O. Anderson, and Douglas Tave) The Haworth Press, Inc., 1993, pp. 9-31. Multiple copies of this article/chapter may be purchased from The Haworth Document Delivery Center [1-800-3-HAWORTH; 9:00 a.m. - 5:00 p.m. (EST)].

9

No cannibalism was observed at any density. Overall, implementation of this plan from 1987 to 1991 increased percid yield (number/ha) four-fold, while increasing average survival by as much as ten-fold, to 60%. Late pond filling and optimization of inorganic fertility levels and stocking densities greatly improved reliability and the economic and ecological efficiency of fish production.

INTRODUCTION

Due to inadequate natural reproduction of fish in suboptimal habitats, many important fisheries are maintained in inland waters through active stocking programs involving the release of millions of fry or fingerlings (Bardach 1976; Lasker 1987). Because the release of larger fish improves survival, culturing fry to the fingerling stage in ponds is a common practice (Bardach 1978; Laarman 1978). Larval fish require live zooplankton of appropriate abundance and size ranges in a habitat with high water quality in order to survive and grow to stockable size (Rosenthal and Hempel 1970; Wong and Ward 1972; Hunter 1981; Raisanen and Applegate 1983; Dabrowski and Bardega 1984; Mills et al. 1989). Maintaining adequate zooplankton abundance while providing high water quality has proved difficult. Success in culturing walleye, *Stizostedion vitreum*, saugeye, walleye ♀ × *S. canadense* ♂, and hybrid striped bass, *Morone chrysops* ♀ × *M. saxatilis* ♂, for stocking in lakes and reservoirs has been shown to be strongly affected by variation in survival in the ponds and in the growth and condition achieved during the culture period (Lynch et al. 1982; Geiger 1983; Geiger et al. 1985; Fox 1989, Fox et al. 1989; Fox and Flowers 1990; Malison et al. 1990; Siegworth and Summerfelt 1990).

Variability in yields makes it difficult to predict how many fingerlings will be available for stocking from year to year. This makes it difficult to comply with management plans and programs. Acreage of ponds, numbers of personnel, fry production facilities, and expenses for fertilizers and other supplies required are all increased by this uncertainty.

Zooplankton availability and water quality are closely related to: the interactions among nutrient concentrations, phytoplankton abundance and species composition, and zooplankton grazing, reproduction, and survival. A pond management regimen that maintains adequate zooplankton for one fish species, therefore, has a high likelihood of working well for others, provided variations in hatchery location and water supply can be taken into account. Review of common hatchery practices showed that most hatcheries added far more fertilizer, particularly phosphate (Culver 1991), than would be suggested from studies of lake eutrophication (Barica et al. 1980). Experiments at Hebron Hatchery showed that algal composition could be manipulated by altering the inorganic N:P ratios in the ponds (Helal 1990) just as had been previously found for lakes (Tilman et al. 1986). Low N:P ratios and infrequent additions of large amounts of fertilizer rather than frequent additions of small amounts favor blue-green algae and filamentous green algae that create inefficiency in food chains and interfere with fish survival, growth, and harvest from ponds. The extent to which variation in fish survival in ponds is influenced by toxic blue-green algae needs further work, but *Oscillatoria* has been shown to have a negative effect on *Daphnia* (Infante and Abella 1985). Recent progress in research on toxic algae such as *Oscillatoria agardhii* (Lindholm et al. 1989; Meriluoto 1989) and *Microcystis aeruginosa* (Eriksson et al. 1988) has identified peptide toxins whose effects on fish are still unknown. High phosphate levels favor large colonial and filamentous algae because of their phosphorus-storage ability (Darley 1982). *Hydrodictyon, Rhizoclonium*, and *Spirogyra* can be particularly troublesome to hatchery managers.

A fertilization and stocking regimen was developed based on: (1) filling ponds immediately before stocking; (2) spraying dilute solutions of liquid inorganic fertilizer; (3) adding no organic fertilizers; (4) fertilizing weekly based on weekly inorganic nitrogen and phosphorus analyses in each pond so that its phosphate concentration was raised to 30 μg PO_4-P/L and Ammonia-N + Nitrate-N concentrations were raised to 600 μg N/L; and (5) stocking sufficient fry to harvest 300,000 fish/ha. For these conditions and species, this meant stocking 450,000 fry/ha.

If a fingerling production strategy such as this is to be useful, it must be equally effective under a variety of types of water supply, pond size and shape and for more than one species of fish. In this research project, therefore, production results for walleye and saugeye were compared and contrasted for three fish hatcheries differing with respect to water supply and pond size and depth, using the new management regimen as compared to previous management regimens used at the same locations.

MATERIALS AND METHODS

Approach

Percid production records for 80 ponds from three hatcheries over 5 years (1986-1990) were made available by the Ohio Department of Natural Resources and were used to determine variation in survival and production (kg/ha) and to compare growth (individual weight at harvest) of walleye vs. saugeye under similar production methods (1991). The number of ponds used for raising walleye and saugeye varied among hatcheries and years, due to changing needs for a given taxon. Fertilization and stocking densities were not sufficiently constant to test yearly differences for significance. The shift to the recommended fertilization regimen was gradual. All hatcheries did use the recommended regimen in 1991, however, so comparison of its efficacy with that of the previous methods as a group was done.

Study Sites

Three different hatcheries in eastern, central, and western Ohio (Figure 1) were used in this study. Each draws water from existing mesotrophic (Senecaville) or eutrophic (Hebron and St. Mary's) reservoirs. Senecaville draws water via a 30-cm diameter pipe from well above the reservoir bottom, and St. Mary's has two similar-sized inputs from near the bottom of the shallow (1.5 m) Grand Lake-St. Mary's Reservoir. Unlike the other two hatcheries, not all St. Mary's ponds can be filled individually, so initial filling and "topping up" often involved flow-through from another pond. Hebron draws water from Buckeye Lake via a 2-km section of the old Ohio canal system accessed through a large inlet pipe. All hatcheries filter the water entering the ponds through 0.5-mm screens to pre-

FIGURE 1. Location of hatcheries used in the study. The distance between Senecaville and St. Mary's is approximately 320 km.

Hatchery locations in Ohio

vent introduction of undesired larval fish. Zooplankton and phytoplankton pass easily through the screens. Ponds were 0.1 to 2.85 ha and contained 1,000 to 59,000 m³ of water. Most averaged 1 m in depth.

Pond Filling and Fertilization Schedules

Previous work (Geiger et al. 1985; Culver 1988) shows that ponds have a profound decline in phyto- and zooplankton abundance in 4 or 5 weeks after filling, with or without fish. This suggests that prompt stocking after filling would be advantageous. Pond records were examined to determine pond filling schedules relative to stocking dates and fertilization patterns. Fertilization costs for 1990 for the liquid, granular, and organic fertilizers were also determined, relative to the number of fish raised. In that year, Senecaville Hatchery personnel added 9.5 kg/ha N as dry ammonium nitrate, 560 kg/ha alfalfa hay at the time of stocking, 56 kg/ha alfalfa meal on each of two occasions, and 4.7 L/ha 10:34:0 ($N:P_2O_5:K_2O$) and 30.0 L/ha 28:0:0 liquid fertilizer. Hebron averaged 8.3 L/ha 10:34:0 and 78.2 L/ha 28:0:0 liquid fertilizer divided into 5 weekly additions. Nitrogen in the 28:0:0 fertilizer was approximately 50% from urea and 50% from ammonium nitrate, while the 10:34:0 fertilizer was ammonium phosphate. Nominal percentages of N and P in fertilizer were not reliable, and required analysis for ammonia, nitrate, urea, and phosphate of appropriately diluted samples from each lot. Urea analyses were performed by local hospital analytical labs on diluted samples using autoanalyzer techniques for blood urea nitrogen. The volume of fertilizer diluted and sprayed into each pond at Hebron (and all hatcheries in 1991) depended upon the nitrate + ammonia and phosphate-P content of each pond (analyses according to methods in APHA et al. [1980]) on the previous day and raised the inorganic N to $600\,\mu g/L$ as N and raised the reactive phosphate to 30 $\mu g/L$ as P each week. At St. Mary's, ponds in the 1990 comparison either received alfalfa meal at 28 kg/ha/week, or 28:0:0 and 10:34:0 in the same manner as at Hebron. Nitrogen and phosphorus content of alfalfa meal is approximately 2.9% and 0.24%, respectively (NRC 1982).

Ponds that were stocked less than 7 days after filling were selected randomly for the fertilizer comparisons, but water was allowed to set for 1 month prior to stocking or fertilization in one set of nine ponds at St. Mary's. In 1991, the three hatcheries all used the same filling schedules and fertilization regimens in a total of 80 ponds–15 for walleye and 65 for saugeye. Fish were usually harvested by draining the water 30 to 40 days after stocking, but some ponds in 1987 were kept in production for as long as 76 days. Fish survival, average individual weight at harvest, and total numerical and weight yields were recorded for all ponds for all years, but growth and yield analyses were restricted to fish harvested 30-50 days after stocking.

Stocking Density Effects

Although counterintuitive, previous experiments had shown that increased stocking densities increased yields both in weight and numbers, because large zooplankton (especially *Daphnia* spp.) overgrazed the available algae in low or no-fish ponds (Culver et al. 1984; Munch et al. 1984; Qin and Culver 1992). Because survival percentage varies greatly among ponds and also because stocking numbers are less precise than harvest numbers, final individual weight and yield as a function of final harvest numbers/ha were used as measures of the effectiveness of the new pond management regimen for producing fish. Most ponds (50/80) were stocked at 400,000 or more fry/ha in 1991, but Hebron Hatchery had a low stocking density (100,000/ha) in 1991 for three walleye ponds and three saugeye ponds as part of additional density-dependence experiments discussed elsewhere (Culver et al. 1992). Because high-density stocking might lead to cannibalism, stomach contents of both walleye and saugeye in a series of ponds stocked at different densities were examined.

Statistical Analyses

Comparison among years, hatcheries, and species for survival was performed with one-way analysis of variance (ANOVA), while yield and individual weight were regressed on harvest density (number/m^2) and were compared using analysis of covariance (ANCOVA) after testing for homogeneity of slopes. A significance level of $\alpha = 0.05$ was set.

RESULTS AND DISCUSSION

Unpredictability of Fish Production

Comparison of survival and growth of both walleye and saugeye from 1986 to 1990 illustrates the wide variation in culturing success and the unpredictability of yields (Table 1). The number of ponds with survival below 10% ("bust ponds") gradually decreased as fertilization levels were reduced to match those found effective at the Hebron hatchery in 1988 (Helal 1990; Helal and Culver 1991). Three of the bust ponds that year at Hebron were experimental ponds with large amounts of fertilizer. The variability in yield shown in Table 1 reflects differences in time between filling and stocking, fertilization treatments, and stocking density; it also indicates the unpredictability of yields. Fish size at harvest was variable from pond to pond also, ranging from 0.17 to 11.07 g wet weight for

TABLE 1. Historical variation in survival and yield (saugeye and walleye combined) for three Ohio hatcheries. Bust ponds are those with less than 10% survival. All ponds, including bust ponds, are included in hatchery-wide average yield estimates. A single asterisk denotes a year when there was partial implementation of the recommended fertilization and stocking plan. Two asterisks denotes years when there was full implementation of the plan.

Year	Bust ponds/ total ponds	Harvest/ha (number)	(kg)	Mean weight (g)
Senecaville				
1989	5/9	72,400	38.3	0.528
1990*	0/8	177,000	45.3	0.256
1991**	1/28	243,600	77.7	0.319
Hebron				
1986	3/25	151,700	55.0	0.363
1987	13/25	69,600	42.1	0.605
1988*	3/29	167,600	59.6	0.355
1989*	0/29	160,500	55.6	0.346
1990**	0/28	230,600	86.7	0.376
1991**	1/28	197,800	68.7	0.347
St. Mary's				
1987	12/22	66,600	103.7	1.555
1988	2/5	47,600	37.2	0.782
1989	11/20	12,800	15.0	1.175
1990*	3/17	62,400	20.0	0.321
1991**	4/24	173,400	57.4	0.331

walleye and from 0.21 to 2.91 g for saugeye. These are unacceptably large pond-to-pond, year-to-year, and hatchery-to-hatchery variations.

Pond Filling Schedule

Ponds that were filled more than 1 week prior to stocking and fertilization had lower survival. The St. Mary's 1990 ponds that were filled one month early had an average survival of 14%, with the lowest survivals occurring in the large (2.5 and 2.8 ha) ponds stocked with saugeye. Average survival for walleye in eight ponds filled within a week before stocking was 64%, with no difference between the ponds fertilized with alfalfa meal and those fertilized with liquid inorganic fertilizer.

Fertilization Regimen

The amounts and timing of fertilizers added to the ponds prior to 1991 varied widely across both hatcheries and time. This contributed to the variation in survival of both walleye and saugeye but the recommended regimen tended to improve survival in both taxa, with the most complete data being available for saugeye (Figure 2). Improvements in yearly survival percentage from early years to 1991 were statistically significant for saugeye at Senecaville ($P < 0.008$), Hebron ($P < 0.01$), and St. Mary's ($P < 0.01$). Statistical comparisons of survival rates were less useful for walleye, since fewer ponds (e.g., 15 of 80 in 1991) were stocked with walleye. As the recommended fertilization regimen was increasingly implemented, the number of bust ponds declined, while yields increased.

Organic fertilization gave variable results and water qualities. In St. Mary's (1990), low organic fertilization alone (25 kg/ha/week alfalfa meal) provided survival and growth comparable to optimal fertilization with inorganic nutrients. At Senecaville in 1989, however, 50 kg/ha/week alfalfa meal combined with inorganics resulted in low survival (3%) when compared to that achieved by inorganic fertilizer only (80%). Dissolved oxygen was lower in organically fertilized ponds (Qin and Culver 1992). No organic fertilizer was used at Senecaville in 1991, and survival and yield were significantly improved (Figure 2 and Table 1).

Cost of Fertilization

The relative costs of inorganic and organic fertilizers were computed in terms of both purchase costs and fish yields. Labor costs were ignored. In 1990, the inorganic treatment at St. Mary's was the most cost efficient (Table 2); both the amounts of fertilizer added and harvest success contributed to the low cost per fish. High nitrate-N concentrations (1,200 μg NO_3^-–N/L) in the water used to fill St. Mary's ponds also decreased the amount of fertilizer required (and hence the cost of fertilizer relative to Hebron). Inorganic nitrogen and phosphate concentrations in incoming waters at Senecaville and Hebron were close to zero. These differences underscore the advantages of measuring N and P concentrations before fertilization with liquid inorganic fertilizers of known composition. Addition of hay

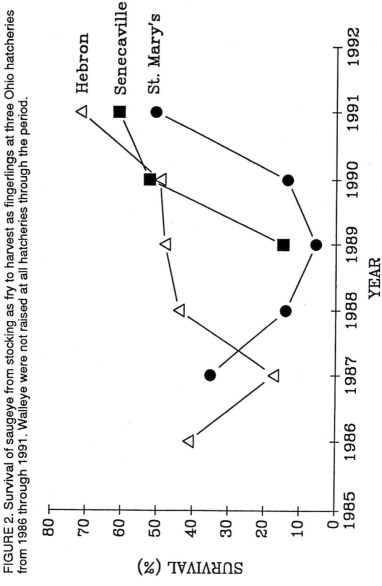

FIGURE 2. Survival of saugeye from stocking as fry to harvest as fingerlings at three Ohio hatcheries from 1986 through 1991. Walleye were not raised at all hatcheries through the period.

△ Hebron
■ Senecaville
● St. Mary's

TABLE 2. Fish yield and relative cost for various fertilization regimens during the 1990 production season at three Ohio Hatcheries. St. Mary's ponds marked with an asterisk were filled one month before stocking and fertilization; all others were filled within one week of stocking.

	No. Ponds	Harvest (No./ha)	Fertilizer costs		
			Inorganic ($/ha)	Organic ($/ha)	Total ($/10³ fish)
Senecaville	8	177,000	24.50	123.75	0.84
Hebron	28	230,600	19.41		0.08
St. Mary's					
Liquid inorganic	4	263,600	10.20		0.04
Organic (alfalfa)	4	232,800		21.60	0.09
Liquid inorganic	9*	43,300	17.76		0.41

and alfalfa meal at Senecaville decreased yield through lower survival, while adding tremendously to cost of materials. Three of the nine St. Mary's ponds filled one month early were deeper than the others listed, so they required more fertilizer to achieve the same nutrient concentrations. The low survival due to early filling greatly increased the cost per fish.

Stocking Density Impacts on Yield

The increased average survival, coupled with the decreased numbers of bust ponds associated with appropriate inorganic fertilization, enabled analysis of the optimal stocking densities for saugeye and walleye. Comparison of yield can provide useful information on the ecological efficiency of management regimens, and yields have increased across all three hatcheries (Table 1). The effect of this management plan on yield cannot be seen directly, however, because yields vary with survival and stocking densities. If yields are regressed on harvest density, however, one can test whether the lines were similar for 1986-1990 (various fertilization regimens and stocking densities) and 1991 (identical fertilization regimens and similar stocking density ranges). Because none of the pairs of regressions was homogeneous with respect to slope ($P < 0.05$), the before-1991 ecological efficiency could not be compared statistically with that of 1991. Yield vs. harvest density relationships among hatcheries for 1991 could, however, be compared, and the slopes were all identical except for walleye yields compared between St. Mary's and Senecaville ($P = 0.012$), so the homogeneous data for all hatcheries were lumped for each species (Figures 3 and 4). Regressions for both saugeye (Figure 3) and walleye (Figure 4) show that increasing the number of fish in the pond increases the yield, up to at least 300,000 fish/ha at harvest.

Given these results, it appears that with these ponds and this management regimen, a harvest density of 300,000/ha represents good success. If 65% survival is a reasonable expectation under favorable conditions, then an appropriate stocking density of larvae should be about 450,000/ha. Not all ponds were in fact stocked at this density in 1991–density experiments at Hebron called for 100,000 fry/ha in six ponds and 500,000 fry/ha in six others. These ponds had been used for hybrid striped bass predation experiments the previous summer and had carry-over growth of *Potamogeton* that interfered with harvest and caused mortality by stranding fish during harvest. These effects are reflected in lower yield and survival at Hebron in 1991. Other ponds had lower stocking densities due to availability of fry.

Nevertheless, the overall improvement in survival and production at the three hatcheries implies that the fertilization and stocking treatments chosen override differences between hatchery characteristics and between walleye and saugeye and that abundance and production of zooplankton are not constant but vary with fish abundance. At low fish densities, the abundance and production of zooplankton, and hence fish yield, are influenced by competitive relationships among zooplankton and overgrazing of algae (Culver et al. 1984, Culver et al. 1992). At higher fish densities, zooplankton abundance and production are controlled more by the reproductive capacity of the zooplankton under high mortality due to fish (Culver et al. 1992). The traditional sense of a carrying capacity does not apply because phytoplankton, zooplankton, and fish biomass all vary profoundly with time. This is not an equilibrium system, because numerical responses by phytoplankton and zooplankton to changing resources and mortality rates are, in turn, influenced by the exponential increase in fish consumption associated with fish growth (Culver et al. 1992).

FIGURE 3. Saugeye yield (wet weight) at three hatcheries in 1991 as a function of numbers harvested. Each point represents a single pond.

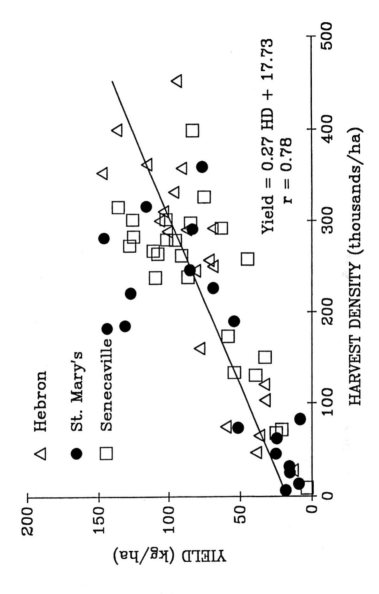

FIGURE 4. Walleye yield (wet weight) at three hatcheries in 1991, as a function of numbers harvested/ha. Each point represents a single pond: SM = Saint Mary's; HB = Hebron; SC = Senecaville.

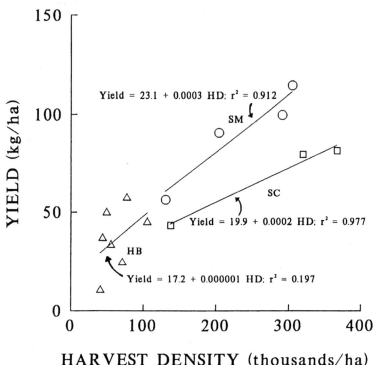

HARVEST DENSITY (thousands/ha)

Stocking Density Impacts on Fish Size at Harvest

Increased survival and yield do not benefit the hatchery manager unless the fish are of sufficient quality to survive once stocked in a reservoir. Survival in reservoirs was outside the scope of this study, but individual weights of fish for both species grown at the three hatcheries were examined. Non-homogeneity of slopes precluded statistical comparison of regressions of individual weight vs. harvest density for early years with those from 1991, but this was not a problem for comparison of regressions for saugeye between Hebron and St. Mary's hatcheries for 1991 (Figure 5). These regressions did not differ signifi-

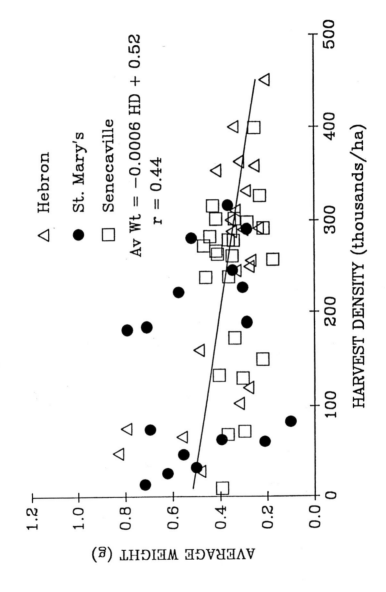

FIGURE 5. Average individual wet weight at harvest for saugeye in 1991 as a function of numbers harvested. Each point represents a single pond.

23

cantly (ANCOVA, $P = 0.699$). Senecaville's regression was not homogeneous with Hebron's, so ANCOVA could not be performed. Nevertheless, a regression line is shown (Figure 5) for all the data combined to demonstrate the important point that there was little difference in individual weight over a range of harvest densities from 200,000 to 400,000 fish/ha, so hatchery managers could increase stocking density within this range without decreasing the size of fish at harvest. Although average weight at harvest was regressed against harvest density, weight actually appeared to decline 50% from a harvest density of essentially no fish up to a harvest density of 100,000/ha and then had a slope of zero up to the highest densities observed (Figure 5). Densities lower than 100,000 fish/ha at harvest produced slightly larger fish at the expense of much lower numbers of fish produced (Figure 5). The research suggests that this is because zooplankton are prevented from overgrazing algae at higher fish densities, so both algae and zooplankton are produced for a longer period before the increased daily ration sizes of fish suppress the zooplankton and the fish switch to eating chironomids (Culver et al. 1992).

Comparative Success for Saugeye and Walleye

Pooling homogeneous data sets for 1991 permits testing whether saugeye and walleye responded differently to these culture methods. Despite the restrictions caused by the small number of ponds used to culture walleye in 1991 (Figure 6), four of the possible six pair-wise comparisons for three hatcheries and two fish species showed no difference between hatcheries, again suggesting that growth relationships are comparable even though water supplies, soil types, and pond shapes and sizes all differed among hatcheries.

Regressions of 1991 yield on harvest density lumped from all hatcheries (Figure 7) showed no significant difference between saugeye and walleye (ANCOVA, $P > 0.328$). The regression of size at harvest as a function of harvest density did not differ significantly (ANCOVA, $P > 0.80$) between the two species (Figure 8). Taken together, these results suggest that the same techniques may be used at all three hatcheries for both species with equivalent results.

FIGURE 6. Average individual wet weight at harvest for walleye in 1991 as a function of numbers harvested: SM = St. Mary's; HB = Hebron; SC = Senecaville.

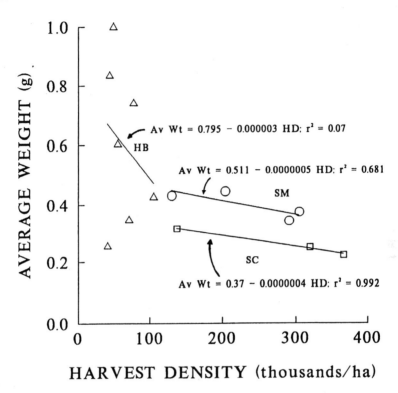

Cannibalism and Fish Stocking Densities

Because cannibalism has been suggested to be responsible for low survival in pond culture, some managers have stocked fewer fish to decrease the encounter rate. During dietary analysis in this study, over one thousand fish stomachs from the three hatcheries were examined. Fish density at harvest ranged from 100,000 to 460,000 fish/ha. Prey found included copepods, cladocerans, and chironomid larvae (Qin and Culver 1992). No cannibalism was observed at any harvest density, so there is no evidence that stocking densities in these ranges caused significant mortality from this source.

FIGURE 7. Saugeye and walleye yields (wet weight) vs. harvest density for all 1991 ponds: circles = saugeye; triangles = walleye.

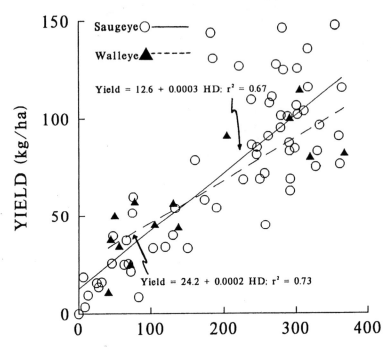

HARVEST DENSITY (thousands/ha)

SUMMARY

This research improved the total yield of percid fingerlings for the three hatcheries four-fold from 1987 to 1991, while increasing survival by as much as ten-fold. The system depended primarily upon photosynthesis by phytoplankton, which in turn supported the zooplankton (Culver et al. 1992). The lack of any organic fertilizer improved water quality by decreasing oxygen consumption (Qin and Culver 1992). Furthermore, use of organic fertilizers such as alfalfa meal would have made it difficult to adjust the relative amounts of nitrogen and phosphorus added to the ponds. Addition of small amounts of inorganic fertilizer each week avoided the

FIGURE 8. Saugeye and walleye average individual wet weights vs. harvest density for all 1991 ponds: circles = saugeye; triangles = walleye.

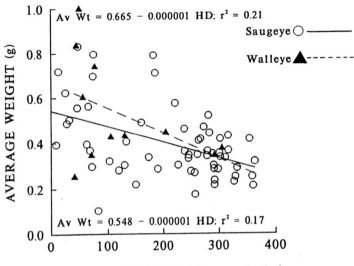

production of large filamentous algae which had been problems during harvest at all three hatcheries. Weekly analyses of inorganic nitrogen and phosphate concentrations assured that appropriate amounts of fertilizer were added. This was particularly important when first filling the ponds, when variation in the nutrient content of water sources could result in initial nutrient concentrations being as much as three times target values. At Senecaville, where high initial nitrate concentrations resulted from agricultural runoff, addition of only phosphate fertilizer encouraged algal growth, and brought the nitrate concentration below the target level within one week. Nutrient determinations were also important at week four or five, when declining algal abundance allowed nitrogen to accumulate. If hatchery managers note the clearing water and add ammonium nitrate and phosphate, ammonia toxicity to fish could result, whereas weekly chemical analysis of pond water will allow addition as needed without over-fertilization. All ponds showed this algal decline at week 4 or 5, so it is important to fill the ponds as soon before adding fry as is practicable. No flushing is required. In fact, flowing water through ponds makes control of nutrient inputs difficult.

Stocking 450,000 fry/ha in ponds appears appropriate for obtaining both the greatest total weight and numbers of fish from the ponds. High stocking densities do not seem to greatly diminish the size of fish. Potentially, even higher yields might be obtained with higher stocking densities.

This study has not considered all the factors that could potentially influence its adoption at other sites. All the hatchery water sources had total alkalinities of 80-111 mg/L as $CaCO_3$ which provided adequate inorganic carbon to support photosynthesis at the fertilizer rates used. pH was not monitored routinely, but values in a few ponds reached 9.7, so pH could be a source of fish mortality on its own and through its influence on free ammonia. The experiments were limited to mid-April through early June, and water temperatures seldom exceeded 23°C. Other fish taxa spawning later in the year would potentially involve a different set of nutrient/phytoplankton/zooplankton interactions. All results were for ponds stocked from productive reservoirs. Hatcheries filling ponds with well water may experience a different set of dynamics for the initiation of the phytoplankton and zooplankton blooms.

ACKNOWLEDGMENTS

We wish to thank D. Uetrecht, M. Giere, L. DePinto, M. Grove, and D. Stoeckel for their very helpful assistance in field and laboratory and also J. Stafford and staff at the Hebron Fish Hatchery, P. Keyes and staff at the Senecaville Fish Hatchery, and R. Chittum and staff at St. Mary's Fish Hatchery for providing all kinds of help and access to experimental ponds. This research was supported by the Ohio Department of Natural Resources Division of Wildlife and the US Fish and Wildlife Service, project F-57-R; the Ohio Cooperative Fish and Wildlife Research Unit; and The Ohio State University.

REFERENCES

APHA et al. (American Public Health Association, American Water Works Association, and Water Pollution Control Federation). 1980. Standard Methods for the Examination of Water and Wastewater, 15th ed. American Public Health Association, Washington, D.C.

Bardach, J. E. 1976. Aquaculture revisited. Journal of the Fisheries Research Board of Canada 33:880-887.

Bardach, J. E. 1978. The growing science of aquaculture. Pages 424-446 *in* S. D. Gerking, ed. Ecology of Freshwater Fish Production. Blackwell Scientific Publications, Oxford.

Barica, J., H. Kling, and J. Gibson. 1980. Experimental manipulation of algal bloom composition by nitrogen addition. Canadian Journal of Fisheries and Aquatic Sciences 37:1175-1183.

Culver, D. A. 1988. Plankton ecology in fish hatchery ponds in Narrandera, NSW, Australia. Verhandlungen Internationale Vereinigung für Theoretische und Angewandte Limnologie 23:1085-1089.

Culver, D. A. 1991. Effects of the N:P ratio in fertilizer for fish hatchery ponds. Verhandlungen Internationale Vereinigung für Theoretische und Angewandte Limnologie 24:1503-1507.

Culver, D. A., R. M. Vaga, and C. S. Munch. 1984. Evidence of size-selective fish predation on the reproductive output of Cladocera in hatchery ponds. Verhandlungen Internationale Vereinigung für Theoretische und Angewandte Limnologie 22:1636-1639.

Culver, D. A., J. Qin, S. P. Madon, and H. A. Helal. 1992. *Daphnia* Production Techniques for Rearing Fingerling Walleye and Saugeye. Completion Report, Federal Aid in Fish Restoration Project F-57-R, for the Ohio Department of Natural Resources and U.S. Fish and Wildlife Service, Columbus, Ohio.

Dabrowski, K., and R. Bardega. 1984. Mouth size and predicted food size preferences of larvae of three cyprinid fish species. Aquaculture 40:41-46.

Darley, W. M. 1982. Algal Biology: A Physiological Approach. Basic Microbiology, Vol. 9, J.F. Wilkins, series ed. Blackwell Scientific Publishers. Oxford.

Eriksson, J. E., J. A. O. Meriluoto, H. P. Kujari, and O. M. Skulberg. 1988. A comparison of toxins isolated from the cyanobacteria *Oscillatoria agardhii* and *Microcystis aeruginosa*. Comparative Biochemistry and Physiology 89C: 207-210.

Fox, M. G. 1989. Effect of prey density and prey size on growth and survival of juvenile walleye *(Stizostedion vitreum vitreum)*. Canadian Journal of Fisheries and Aquatic Sciences 46:1323-1328.

Fox, M. G., and D. D. Flowers. 1990. Effect of fish density on growth, survival, and food consumption by juvenile walleye in rearing ponds. Transactions of the American Fisheries Society 119:112-121.

Fox, M. G., J. A. Keast, and R. J. Swainson. 1989. The effect of fertilization regime on juvenile walleye growth and prey utilization in rearing ponds. Environmental Biology of Fishes 26:129-142.

Geiger, J. G. 1983. A review of pond zooplankton production and fertilization for the culture of larval and fingerling striped bass. Aquaculture 35:353-369.

Geiger, J. G., C. J. Turner, K. Fitzmayer, and W. C. Nicols. 1985. Feeding habits of larval and fingerling striped bass and zooplankton dynamics in fertilized rearing ponds. Progressive Fish-Culturist 47:213-223.

Helal, H. A. 1990. Nitrogen:Phosphorus Ratio and Trophic Dynamics in Fish-hatchery Ponds. Doctoral dissertation, The Ohio State University, Ohio.

Helal, H. A., and D. A. Culver. 1991. N:P ratio and plankton production in fish ponds. Verhandlungen Internationale Vereinigung für Theoretische und Angewandte Limnologie 24:1508-1511.

Hunter, J. R. 1981. Feeding ecology of marine fish larvae. Pages 33-77 *in* R. Lasker, ed. Marine Fish Larvae. Washington Sea Grant Program, University of Washington, Seattle, Washington.

Infante, A., and S. E. B. Abella. 1985. Inhibition of *Daphnia* by *Oscillatoria* in Lake Washington. Limnology and Oceanography 30:1046-1052.

Laarman, P. N. 1978. Case histories of stocking walleyes in inland lakes, impoundments and the Great Lakes–100 years of walleye. Pages 254-260 *in* R. L. Kendall, ed. Selected Coolwater Fishes of North America. Special Publication No. 11, American Fisheries Society, Washington, D. C.

Lasker, R. 1987. Use of fish eggs and larvae in probing some major problems in fisheries and aquaculture. Transactions of the American Fisheries Society Symposium 2:1-16.

Li, S., and J. A. Mathias. 1982. Causes of high mortality among cultured larval walleyes. Transactions of the American Fisheries Society 111:710-721.

Lindholm, T., J. E. Eriksson, and J. A. O. Meriluoto. 1989. Toxic cyanobacteria and water quality problems–Examples from a eutrophic lake on Åland, South West Finland. Water Research 23(4): 481-486.

Lynch, W. E., B. L. Johnson, and S. A. Schell. 1982. Survival, growth and food habits of walleye × sauger hybrids (saugeye) in ponds. North American Journal of Fisheries Management 4:381-387.

Malison, J. A., T. B. Kayes, J. A. Held, and C. H. Amundson. 1990. Comparative survival, growth, and reproductive development of juvenile walleye and sauger and their hybrids reared under intensive culture conditions. Progressive Fish-Culturist 52:73-82.

Meriluoto, J. A. O., A. Sandström, J. E. Eriksson, G. Remaud, A. Grey Craig, and J. Chattopadhyaya. 1989. Structure and toxicity of a peptide hepatotoxin from the cyanobacterium *Oscillatoria agardhii*. Toxicon 27:1021-1034.

Mills, E. L., M. V. Pol, R. E. Sherman, and T. B. Culver. 1989. Interrelationships between prey body size and growth of age-0 yellow perch. Transactions of the American Fisheries Society 118:1-10.

Munch, C. S., R. M. Vaga, and D. A. Culver. 1984. Evidence for size-selective grazing of phytoplankton species by zooplankton in fish hatchery ponds. Verhandlungen Internationale Vereinigung für Theoretische und Angewandte Limnologie 22:1640-1644.

NRC (National Research Council [U.S.], Subcommittee on Feed Composition.) 1982. United States-Canadian Tables of Feed Composition: Nutritional Data for United States and Canadian feeds, 3rd revision. National Academy Press, Washington, D.C.

Qin, J., and D. A. Culver. 1992. The survival and growth of larval walleye,

Stizostedion vitreum, and trophic dynamics in fertilized ponds. Aquaculture 108:257-276.

Raisanen, G. A., and R. L. Applegate. 1983. Prey selection of walleye fry in an experimental system. Progressive Fish-Culturist 45:209-214.

Rosenthal, H., and G. Hempel. 1970. Experimental studies in feeding and food requirements of herring larvae *(Clupea harengus L.)*. Pages 344-364 *in* J. H. Steele, ed. Marine Food Chains. Oliver and Boyd, Edinburgh.

Siegworth, G. L., and R. C. Summerfelt. 1990. Growth comparison between fingerling walleyes and walleye × sauger hybrids reared in intensive culture. Progressive Fish-Culturist 52:100-104.

Tilman, D., R. Kiesling, R. Sterner, S. S. Kilham, and F. A. Johnson. 1986. Green, blue-green and diatom algae: Taxonomic differences in competitive ability for phosphorus, silicon and nitrogen. Archiv für Hydrobiologie 106:473-485.

Wong, B., and F. J. Ward. 1972. Size selection of *Daphnia pulicaria* by yellow perch *(Perca flavescens)* fry in West Blue Lake, Manitoba. Journal of the Fisheries Research Board of Canada 29:1761-1764.

Pond Production of Fingerling Walleye, *Stizostedion vitreum,* in the Northern Great Plains

Robert C. Summerfelt
Christopher P. Clouse
Lloyd M. Harding

ABSTRACT. The cultural practices used to produce fingerling wall-eye, *Stizostedion vitreum,* in drainable earthen ponds are described for a state fish hatchery in Nebraska and two federal hatcheries in North Dakota. The ponds were filled 1 to 7 days before D2-D4 (D1=the day of hatch) walleye fry were stocked. At one hatchery, ponds were sometimes double-cropped, first for production of northern pike, *Esox lucius.* The two federal hatcheries fertilized ponds with ground alfalfa hay or pellets, while the standard practice at the Nebraska hatchery was not to fertilize walleye ponds, because of concern that fertilization would result in weed problems and oxygen depletion. One hatchery seeded the ponds with rye grass in the fall. Two of the hatcheries regularly used herbicides to prevent the stranding of fingerlings during harvest and their mortality caused by entangment with net algae, *Hydrodicton.* When used, herbicide

Robert C. Summerfelt, Christopher P. Clouse, and Lloyd M. Harding, Department of Animal Ecology, Iowa State University, Ames, Iowa 50011-3221, USA.

Correspondence for Christopher P. Clouse may be addressed to Milford Fish Hatchery, RR 3, Box 304BB, Junction City, KS 66441, USA.

Correspondence for Lloyd M. Harding may be addressed to Coldwater Fish Farms, Inc., P.O. Box 1, Lisco, NE 69148, USA.

[Haworth co-indexing entry note]: "Pond Production of Fingerling Walleye, *Stizostedion vitreum,* in the Northern Great Plains." Summerfelt, Robert C., Christopher P. Clouse, and Lloyd M. Harding. Co-published simultaneously in the *Journal of Applied Aquaculture,* (The Haworth Press, Inc.) Vol. 2, No. 3/4, 1993, pp. 33-58; and: *Strategies and Tactics for Management of Fertilized Hatchery Ponds* (ed: Richard O. Anderson and Douglas Tave) The Haworth Press, Inc., 1993, pp. 33-58. Multiple copies of this article/chapter may be purchased from The Haworth Document Delivery Center [1-800-3-HA-WORTH; 9:00 a.m. - 5:00 p.m. (EST)].

treatment was applied before ponds were filled (Aquazine™) or as needed during the culture interval (Aquazine™ or copper sulfate). Harvesting was done after 24 to 58 days; the extreme range represented variation among hatcheries; the variation among ponds at a given hatchery ranged from 4 to 10 days. Harvest occurred when fingerlings were 25 to 50 mm total length and weighed 1,500-5,440 fish/kg. Harvests ranged from 11,933 to 308,537 fingerlings/ha. Survival ranged from 3 to 104% of the estimated number of fry stocked.

INTRODUCTION

Yellow perch, *Perca flavescens,* and walleye, *Stizostedion vitreum,* are the most exploited percid species in North American commercial and recreational fisheries (Kendall 1978). To sustain walleye fisheries, state, federal, and provincial fisheries agencies in North America stocked more than one billion walleye fry and fingerlings in 1984 (Conover 1986). Numerically, fry account for 98% of the walleye stockings in the United States and Canada, but the relative survival of fingerling walleye is often much greater than that of fry (Heidinger et al. 1985; Paragamian and Kingery 1992).

Nearly all fingerling walleye are raised in ponds. The cultural process involves capturing and spawning wild broodstock, incubating eggs, and stocking and managing ponds. Fingerling culture generally has been limited to production of 35- to 50-mm total length (TL) fingerlings, because of cannibalism and loss of zooplankton bloom (Colesante et al. 1986). The ability to produce walleye fingerlings in large quantities is, perhaps, the most important problem in the culture of coolwater fishes today (Nickum 1978). Reliable techniques for mass culture of walleye fry raised on formulated feed in intensive culture environments have not developed to date. Non-inflation of the gas bladder (failure of the gas bladder to inflate) has been suggested as the key factor in poor survival of tank-raised fry fed formulated feed (Colesante et al. 1986; Barrows et al. 1988). Therefore, at this time, pond culture is the only practical method to raise the large number of fingerlings that are needed for management.

Production of walleye fingerlings in ponds is dependent on pond management activities that are effective in developing and sustaining adequate densities of microcrustacean zooplankters and benthic chironomids, which are used by the walleye as their principle food until they attain or exceed 30 mm TL (Merna 1977; Mathias and Li 1982; Raisanen and Applegate 1983; Fox et al. 1989). Desirable survival and growth of walleye in ponds also requires minimal numbers of undesirable animal species (e.g., organisms that prey on fish or compete with fish for food) and adequate dissolved oxygen. Control of filamentous algae or macrophytes is necessary to avoid entrapment of fish during draining or seining (Nickum 1986). Survival of fry to fingerlings is still unpredictable and extremely variable (Beyerle 1979).

Procedures to culture walleye fry to a desirable sized fingerling in earthen ponds have not been standardized, and culture methods are site-specific. A high fry stocking density is desired in both the public and private sectors to optimize capital costs of pond construction. High fry stocking density demands use of intensive pond management strategies; i.e., methods that will enhance production of desirable invertebrate food items, usually focused on management of the zooplankton population (Richard and Hynes 1986). Well-timed filling of ponds, zooplankton seeding (inoculation of ponds with cladocera), and pond fertilization have produced and maintained large numbers of desirable zooplankton in drainable/undrainable, natural, and plastic-lined ponds (Kuss 1988).

Organic fertilizers are commonly used in walleye fingerling production to produce the zooplankton forage base for the fish (Dobie 1956; Richard and Hynes 1986; Fox and Flowers 1990). Often, however, a diverse complex of edaphic factors (soil type, water fertility, time between filling, and stocking) influences the abundance and composition of the zooplankton community in ways that are still unpredictable. Production may be higher and less variable with fertilization, but it increases the biochemical oxygen demand, and this can lead to depletion of dissolved oxygen and to fish kills.

The northern Great Plains is a major walleye production region; 73% of all fingerling walleyes produced by the U.S. Fish and Wildlife Service are raised in ponds at the coolwater fish hatcheries located in North Dakota (Valley City National Fish Hatchery and Garrison Dam National Fish Hatchery) and South Dakota (Gavins Point National Fish Hatchery) (FWS 1991). In 1984, the states of Minnesota, Iowa, Wisconsin, and South Dakota produced 58% of

all fry and 92.2% of all fingerlings stocked by the 10 leading–
ranked by size of their walleye stocking activities–state and provincial agencies (Conover 1986). In 1990, ninety percent of the 20.2
million walleye stocked by the Nebraska Game and Parks Commission were produced at the North Platte State Fish Hatchery, one of
the three hatcheries studied in this project.

The objective of this paper is to describe the practices used for
pond production of phase I fingerlings (30- to 60-mm) by three
hatcheries in the northern Great Plains (Table 1). These case histories are given as examples of contemporary culture techniques for
this species. The study sites are the Nebraska's North Platte State
Fish Hatchery and two U.S. Fish and Wildlife Service fish hatcheries in North Dakota–Valley City National Fish Hatchery and Garrison Dam National Fish Hatchery. Each hatchery manages its ponds
based on factors related to local circumstances, but the management
strategies used at these hatcheries represent contemporary practices
for walleye fingerling culture in drainable earthen ponds. This information should be valuable reference data for comparative purposes.
Walleye production in earthen ponds differs slightly from practices
used for production of walleye in plastic-lined ponds (Jahn et al.
1989), but they are substantially different from practices used for
raising walleye fingerlings in undrainable pond environments
where harvest is accomplished by using fyke nets (Gustafson 1991).

NORTH PLATTE STATE FISH HATCHERY

Study Site

The North Platte State Fish Hatchery (NPSFH), located 5 km
south of North Platte, Nebraska is used for warm- and coolwater
fish culture. The species include walleye, northern pike, *Esox lucius,*
channel catfish, *Ictalurus punctatus,* and yellow perch. In the past,
eggs have been taken from broodstock collected from various Nebraska reservoirs, but in 1990 and 1991 eggs were taken from brood fish
in Merritt Reservoir in north central Nebraska.

This hatchery has thirty-eight 0.25-ha ponds, four 0.125-ha
ponds, and one 0.5-ha pond. All ponds have a mean depth of

TABLE 1. Pond management practices used at the Garrison Dam National Fish Hatchery (GDNFH), at the Bald Hill Unit of Valley City National Fish Hatchery (VCNFH), and at the North Platte, Nebraska State Fish Hatchery (NPSFH).

Parameter	GDNFH (1989-90)	VCNFH (1989-90)	NPSFH (1990-91)
Water source	Lake Sakakawea (1989) and Riverdale Spillway Lake (1990)	Lake Ashtabula	NPPD canal from Lake Maloney
Pond size (ha)	0.6	0.3	0.4
Filling	middle to late May (after northern pike fingerling production)	early May	early to middle April (7-10 days pre-stock)
Fertilizer used	alfalfa hay	alfalfa pellets	none
Fertilizer application rate	1,600 kg/ha	700 kg/ha	none
Stocking dates	late May, early June	early May	middle April
Stocking density (No./ha)	390,000	375,000	250,000
Harvest dates	late June, early July	middle June	late May, early June
Mean water temperature (°C)	20	17	15
Range	15.0-24.2	11.2-22.4	9-22
Aquazine™	not used	as pre-emergent (1.12 kg/ha), during culture (1.7 - 2.0 ppm)	as needed (2.2 kg/ha)
Copper sulfate	occasionally used to control algae (0.18 - 0.23 ppm)	occasionally used to control algae (0.40 ppm)	not used
Pond preparation	ryegrass is planted each year after draining	ponds are disked in the spring	ponds are disked in the spring

approximately 1 m. Each pond is equipped with a concrete outlet control structure (chimney or monk) at the deep end, which leads to an outside concrete catch basin where the fish are collected when the pond is drained.

Pond Preparation

The ponds were drained and disked in the fall and kept dry until they were refilled for the walleye culture season, which generally began in mid-April (Table 1). Filling the ponds generally required about 3 days and began about one week before fry were stocked. The number of ponds used each year was based on management requests, but generally about 15 of the 0.25-ha ponds were used for production of phase I walleye fingerlings. The ponds were not fertilized; aquatic insect control measures (oil) were not used; and herbicides were used only as needed.

Water Supply

The water supply for the NPSFH was drawn from the Nebraska Public Power District's (NPPD) hydroelectric/irrigation canal. Water temperatures during the culture period were 8 to 10°C when the fry were stocked, but temperature reached 18° to 22°C at harvest, 40 to 60 days later. Rotifers, copepods, and cladocerans were present in the inflow water, with total densities ranging from 100 to 800 zooplankters/L.

Stocking and Management

D2 walleye fry for stocking were enumerated by volumetric displacement (Piper et al. 1982) and were stocked in the ponds in early morning at a rate of 250,000/ha. Because of seepage and evaporation, water was added continually throughout the culture period to maintain the desired level in the ponds. The rate of inflow was equivalent to one water exchange every 14 days. The ponds were not fertilized prior to experimental studies in 1991, but the productive capacity probably was augmented by inorganic nutrients and zooplankton in the water supply. Zooplankton densities in the ponds during the phase I production (fry to fingerlings) ranged from 500 to 2,500 zooplankton/L, usually composed of cyclopoid and calanoid copepods, the cladoceran genera *Daphnia* and *Bosmina,* and an abundance of rotifers. Chironomid larvae, the major food of walleye in the last two weeks of culture, were abundant in the pond substrate, but their densities were not enumerated.

Rooted aquatic vegetation, such as *Chara* and small pond weed, *Potamogeton,* was a problem. During the phase I culture period, copper sulfate or Aquazine™[1] was applied in liquid form as needed to the pond surface at rates of 0.5 and 2.2 mg/L, respectively, by using a broadcast sprayer. Herbicides were used to avoid trapping the fish in the vegetation at harvest. In some years, 70- to 150-mm TL triploid grass carp, *Ctenopharyngodon idella,* were stocked (500 to 1,500/ha) with walleye to control macrophytes in the phase II production.

Harvest Methods

Harvest began in late May or early June, 40 to 60 days after stocking, when fingerlings reached 40 to 60 mm. Individual ponds required 2 to 3 days to drain, but three or four ponds were often drained simultaneously. Generally, it required 9 to 10 days to harvest all of the ponds. Samples of fish were counted and weighed to determine the number/kg, from which total number of fingerlings harvested was estimated. Fingerlings were held in indoor tanks until they were stocked into area lakes, which was usually within 24-48 hours after harvest.

Three to 4 days after harvest of the phase I fingerlings, the ponds were refilled and restocked (25,000/ha) with fingerling walleyes that were cultured for another 90 days. This phase II production was used to produce a fingerling that reached lengths of 100 to 150 mm by September. This paper discusses only the phase I culture interval.

Harvest Results

The average yield of phase I fingerlings for a 19-year interval (1973-1991) was 66.5 ± 5.4 kg/ha (mean ± SE), and ranged from

---

1. Use of trade or manufacturer names does not imply endorsement.

TABLE 2. Mean annual survival and yield and coefficients of variation (CV) of fingerling walleye production from culture ponds at the North Platte State Fish Hatchery for the 19 year interval, 1973-1991.

Year	Ponds (number)	Survival Mean (%)	CV	Yield Mean (kg/ha)	CV
1973	6	63.3	24.1	88.5	18.9
1974	5	81.9	15.9	69.2	11.1
1975	5	42.0	18.9	58.4	10.0
1976	8	25.5	46.2	32.7	40.6
1977	8	31.7	41.0	35.2	37.5
1978	6	45.1	31.8	57.4	34.3
1979	6	73.6	11.8	57.6	19.3
1980	7	67.5	21.7	83.3	29.9
1981	6	73.1	31.3	67.5	22.3
1982	13	56.7	43.5	48.8	71.9
1983	16	65.0	45.8	62.0	41.0
1984	15	54.6	43.0	55.3	44.5
1985	17	86.9	13.5	101.2	22.7
1986	16	41.8	65.3	40.8	43.6
1987	16	104.2[a]	11.0	138.9	20.5
1988	13	57.5	26.6	68.3	42.8
1989	14	22.2	28.1	50.5	26.4
1990	17	104.9[a]	13.9	93.3	19.9
1991	17	60.4	26.9	54.1	25.5
Mean		60.9	29.5	66.5	30.7
		38.7[b]		38.6[b]	

[a] Survival values greater than 100% reflect error in the volumetric procedure used to estimate number of fry at stocking.
[b] CV of annual means.

32.7 to 138.9 kg/ha (Table 2). The coefficient of variation (CV) in mean yield for the 19 yearly means was 38.6%. The mean CV for yield within each year, which is an expression of the variation among the 5-17 ponds per year, was $30.7 \pm 3.4\%$. Annual mean survival was $60.9 \pm 5.4\%$, with a CV of 38.7%, basically identical to the CV for yield. There was a strong linear correlation ($r = 0.83$; $P < 0.001$) between yield and survival (Figure 1). The average CV for within-year variability in survival (i.e., pond-to-pond variability) was $29.5 \pm 3.4\%$.

In 1990-91, at the NPSFH, survival was 74%, yield was 74.8 kg/ha, and 199,796, thirty- to 40-mm fingerlings/ha were harvested (Table 3).

Prey and Walleye Food

In 1990 and 1991, walleye fed on copepods, cladocerans, and chironomid larvae and pupae, but not on rotifers. The smaller cope-

FIGURE 1. Relationship between mean annual yield (kg/ha) and survival (%) of walleye at the North Platte State Fish Hatchery from 1973-91 (Data from Tom Ellis, North Platte State Fish Hatchery, pers. comm.).

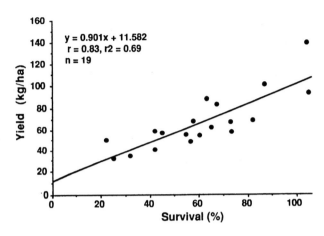

pods and cladocerans were generally preferred early in the culture season, and food preference gradually shifted to larger chironomids and cladocerans in the latter stages of the culture period. Zooplankton densities usually were well above 100/L when the larvae began feeding. Zooplankton densities declined during a 10-day interval of cold weather in 1990 and after herbicide ($CuSO_4$) application in 1991.

VALLEY CITY NATIONAL FISH HATCHERY

Study Site

Valley City National Fish Hatchery (VCNFH) is located approximately 1 km north of Valley City, North Dakota. The studies in 1989 and 1990 were carried out at the Bald Hill Unit of the hatchery located below Lake Ashtabula. The Bald Hill Unit consists of twelve 0.30-ha earthen ponds with an average depth of 1 m. Each pond was equipped with a concrete kettle at the deep end, which allows adjustment in water level and serves to drain pond water and fish to an outside catch basin.

TABLE 3. Harvest data (range in parentheses) for the Garrison Dam National Fish Hatchery (GDNFH), the Bald Hill Unit of Valley City National Fish Hatchery (VCNFH), and the North Platte State Fish Hatchery (NPSFH).

	GDNFH (1989-90)	VCNFH (1989-90)	NPSFH (1990-91)
Stocking rate (No./ha)	390,000	375,000	250,000
Culture Interval (days)	28 (24-29)	45 (39-52)	50 (45-59)
Mean length (mm)	30 (27-34)	32 (25-40)	45 (30-40)
Mean weight (g)	0.33 (0.24-0.52)	0.28 (0.25-0.36)	0.35 (0.18-0.54)
Number of fish/kg	3,361 (2,425-2,866)	3,904 (1,985-4,400)	2,742 (2,339-3,141)
Condition factor (K)	0.938 (0.715-1.665)	0.868 (0.338-1.336)	0.659 (0.587-0.709)
Yield (kg/ha)	47.9 (6.3 - 94.4)	45.4 (14.7-76.5)	74.8 (46.7-113.2)
No. harvested/ha	197,825 (11,942-322,365)	213,038 (60,929-378,080)	199,796 (146,568-266,538)
Survival (%)	50 (3-81)	56 (15-96)	74 (48-100)

Pond Preparation

Ponds were disked and treated with a pre-emergent application of Aquazine™ to control aquatic vegetation before they were filled (Table 1). Walleye eggs were obtained in mid-April from brood fish collected from Devils Lake, North Dakota. Pond filling was initiated approximately 10 days before the eggs began to hatch and filling generally required 2 days per pond.

Water Supply

The ponds were filled with water drawn from Lake Ashtabula. The water supply had a high buffering capacity, with total alkalinity of 295 mg/L as $CaCO_3$. The pH of the water was generally between 8.2 and 8.6. Water temperatures at the beginning of culture were near 12°C and generally increased about 1.5°C per week during the 5- to

6-week culture interval. Water temperatures at harvest averaged 21°C.

The inflow water to fill the ponds was run through nylon filters (5.0 mm openings) placed over the end of the inflow pipes to prevent wild fish from being introduced. The water passing through the nylon filter contained substantial numbers of zooplankton; large (>2 mm) *Daphnia* sp. which were too large to be consumed by first feeding larvae made up 46% of the total count. Cladocera and copepoda densities in the water supply from Lake Ashtabula in 1990 averaged 129/L throughout the culture interval.

Stocking and Management

The ponds were stocked with D2-D4 fry at a density of 375,000/ha. Ponds were fertilized with alfalfa pellets as the ponds were filled. A total of 800 kg/ha of pellets were applied in four weekly applications during the culture period. The pellets were applied from the pond levees with a blower. Fertilizer application was managed to prevent excessive algal blooms. When dissolved oxygen levels were less than 3 mg/L, inflow water was added to the ponds, and a portable paddle wheel aerator was used to increase oxygen concentrations. Zooplankton populations were monitored by visual inspections of zooplankton densities sampled with a plankton net. Fertilizer application schedules were timed to encourage maximum zooplankton blooms.

Herbicides were used to control problem aquatic vegetation. Net algae, *Hydrodicton,* was the major problem. In 1989, Aquazine™ (80% simazine) was sprayed on the pond bottoms as a pre-emergent herbicide one month before filling at 1.12 kg/ha. A second application of 2.0 mg/L was required in six of the ponds on the 21st day of culture to control an outbreak of net algae. In 1990, Aquazine™ was not applied as a pre-emergent herbicide; copper sulfate was applied to all ponds during the second week of culture at a rate of 0.4 mg/L, and Aquazine™ was used in the fourth week of culture at a rate of 1.7 mg/L to control net algae growth.

Harvest Methods

Harvesting was done by draining the water to an outside catch basin. The catch basins were fitted with dam boards and screens with openings of approximately 3 mm. Fish were weighed as they were removed, and the total number harvested was estimated by making three to four sample counts of the number of fingerlings/kg and by multiplying the average number/kg by the total weight removed.

Harvest Results

Fingerlings were harvested in middle to late June (39 to 52 days after stocking), when they were about 32 mm TL. Average yield was 50.1 kg/ha in 1989 and 40.7 kg/ha in 1990. Survival averaged 71% in 1989 and 42% in 1990 (Table 3).

Prey Items

Zooplankton populations were dominated by cladocera (Figure 2). Chironomid larvae were abundant in the pond sediments and were also are important food items for the fingerlings. Prey selection progressed from small copepods (<1.0 mm) to larger cladocerans (up to 2.0 mm) to chironomid larvae.

GARRISON DAM NATIONAL FISH HATCHERY

Study Site

Garrison Dam National Fish Hatchery (GDNFH) is located in west-central North Dakota, approximately 100 km north of Bismark, North Dakota. Hatchery ponds and buildings are located directly below the dam of Lake Sakakawea, a 306,000-ha impoundment of the Missouri River. The ponds averaged 0.60 ha and had a mean depth of 1 m. In 1989, forty new ponds were added, giving the hatchery a total of 64 ponds for producing northern pike, walleye, or a double-crop of northern pike followed by walleye. The hatchery also has indoor culture facilities for salmonids.

The walleye culture season at GDNFH began after the northern pike were harvested. However, there were many logistics problems involved in managing 64 ponds, especially in regard to timing the incubation of walleye eggs in order to have fry available to stock

FIGURE 2. Mean weekly prey densities (copepoda, cladocera, and chironomids) at Valley City National Fish Hatchery (VCNFH) and Garrison Dam National Fish Hatchery (GDNFH) in 1989-90.

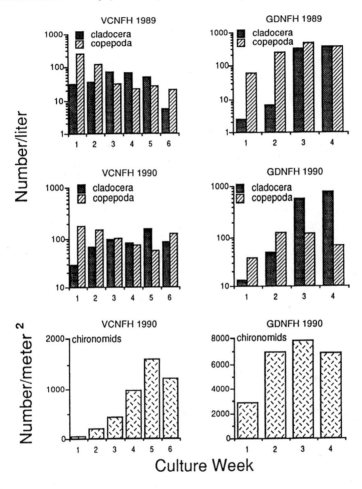

ponds immediately after northern pike were harvested. Fry must be stocked within 3 days after hatching. Northern pike culture began in early April and ended in mid- to late May when the fingerlings were harvested. Normally, this hatchery obtained walleye eggs from brood fish collected from Devils Lake, North Dakota, but because these eggs were available earlier than desired if hatched at the normal incubation temperature (13-15°C), the incubation period was often extended to delay hatching of walleye until the northern pike were harvested.

Pond Preparation

Pond filling was initiated in late May and early June, approximately 10 days before the eggs began to hatch. Each pond generally required two days to fill, but valves were situated so several ponds could be filled simultaneously.

Water Supply

The older ponds were filled with water drawn from Lake Sakakawea (1989 data), but the new ponds were filled with water drawn from the Riverdale Spillway Lake (1990 data). Inflow pipes were situated at the deep end of the ponds, as were the concrete drain basins that drained to outside catch basins. The water supply had a total hardness of 210 to 231 mg/L as $CaCO_3$; the pH of the water was generally between 8.2 and 8.6. Pond water temperatures at the beginning of culture were approximately 15°C and gradually increased to 25°C during the 4-week culture interval. Water temperatures fluctuated with changes in air temperature but generally rose about 2.5°C per week.

Stocking and Management

Walleye fry were held in indoor culture tanks for up to 3 days before stocking. The fry were enumerated by using the water displacement method. D2-D4 fry were stocked at 394,000/ha in the early morning on the side of the pond protected from strong wave action.

Other than filling and maintaining water levels, the only pond management practice was fertilization. When the ponds were filled for northern pike, the rye grass that had been planted the previous fall to protect the pond banks against erosion was flooded; the rye grass may have provided a surface for adult chironomids to lay their eggs. The decomposition of the rye grass provided organic matter in the double-crop ponds, but most of the benefits of the rye grass were probably lost when the ponds were drained to harvest the northern pike. Fertilization of the walleye ponds was done with ground alfalfa hay; the first application was added to the ponds as they were filling. Four hundred kg/ha of hay was applied weekly during the culture period in four equal applications (1,600 kg/ha). The hay was added to the ponds with a dump truck at a location where wind and wave action would distribute the hay throughout the pond.

The ponds used at GDNFH did not have many problems with vegetation, and the only herbicide used in 1989-90 was copper sulfate; it was applied to two ponds in 1989, two days prior to the walleye harvest.

Harvest Methods

Fingerlings were harvested by draining the ponds 24 to 29 days after stocking (late June to early July), when fish were 32 ± 2 mm. Ponds were drained in about 24 hours into an outside catch basin. The catch basins were fitted with dam boards and screens with 3 mm openings.

Harvest Results

Survival averaged 49.8% in 1989 and 51% in 1990 (CV = 28.9%). Yield averaged 38.2 kg/ha in 1989 and 57.6 kg/ha in 1990 (Table 3).

Prey Items

In 1989, zooplankton counts were made weekly from the ponds that were supplied with water from Lake Sakakawea. The zooplankton in the water supply, which exceeded 400 zooplankters/L, was dominated by cyclopoid copepods (Table 4). This zooplankton may have served as an innoculum for these ponds, as well as a first food for walleye. Zooplankton populations in the ponds normally consisted of cyclopoid and calanoid copepods, rotifers, and cladocerans. Chironomids were abundant in the dredge samples of the pond substrates at this hatchery; chironomids averaged nearly $3,000/m^2$ in week 1 to nearly $8,000/m^2$ in week 4 (Figure 2). The

TABLE 4. Mean zooplankton (no./L) density in the water supply (Lake Ashtabula and Lake Sakakawea) used to fill ponds at Valley City National Fish Hatchery and Garrison Dam National Fish Hatchery.

Organism	Lake Ashtabula (VCNFH)[a]		Lake Sakakawea (GDNFH)[b]	
	No./l	%	No./l	%
Cyclopoid copepods	18.3	14.2	306.2	73.4
Calanoid copepods	11.0	8.5	5.5	1.3
Copepod nauplii	23.8	18.5	21.7	5.2
Daphnia sp.	59.7	46.4	36.0	8.6
Bosmina sp.	0.6	0.5	45.0	10.8
Chydorus sp.	1.6	1.3	0.0	0.0
Rotifers	13.7	10.6	2.7	0.7
Total	128.7	100.0	417.1	100.0

[a] Values represent means of 20 daily waters samples collected in May, 1990 from the main inlet pipe flowing into ponds at VCNFH from Lake Ashtabula.

[b] Values represent means of four daily water samples collected in June, 1990 from the main inlet pipe flowing into ponds at GDNFH from Lake Sakakawea.

zooplankton population in water drawn from the Riverdale Spillway Lake, which was used to supply water to the ponds studied in 1990, was not analyzed.

EFFECT OF LENGTH OF THE CULTURE PERIOD ON PRODUCTION VARIABLES

At all hatcheries, all ponds were stocked on the same day; however at harvest, only two to four ponds were harvested each day, which resulted in 4- to 10-day harvest intervals: 9 to 10 days at NPSFH; 4 to 10 days at VCNFH; 4 to 5 days at GDNFH. Because not all ponds were harvested on the same day, there were differences in the lengths of the culture periods among ponds. There was a significant inverse relationship in the length of the culture interval and fingerling survival at NPSFH in 1991 (Table 5); the regression of survival on length of the culture interval ($s = 189.5 - 2.33X$) indicated that survival was reduced 2.3%/day over the 10-day range

TABLE 5. Relationships (correlation coefficient, r) between length of the culture interval (days) and survival (%), mean length (mm), and mean weight (g) of walleye fingerlings at the North Platte State Fish Hatchery (NPSFH), Valley City National Fish Hatchery (VCNFH), and Garrison Dam National Fish Hatchery (GDNFH).

Hatchery Year	Survival r	Survival P	Length r	Length P	Weight r	Weight P
NPSFH						
1990	-0.41	0.39	0.82	0.01	0.81	0.02
1991	-0.72	0.01	0.71	0.01	0.66	0.01
VCNFH						
1989	0.11	0.62	-0.60	0.17	-0.54	0.57
1990	-0.53	0.10	0.74	0.04	0.78	0.05
GDNFH						
1989	-0.41	0.26	0.76	0.25	0.83	0.05
1990	0.22	0.94	0.14	0.60	0.14	0.35

in harvest interval. However, the correlation for the survival-culture interval was not significant at NPSFH in 1990 or at the other hatcheries (Table 5). There was a significant positive correlation between average fish length at harvest and length of the culture interval at NPSFH in both 1990 and 1991 and at VCNFH in 1990; there was also a significant positive correlation with weight in four of the six hatchery-year analyses (Table 5).

DISCUSSION

The phase I walleye fingerling production practices used at the three hatcheries in the study group differed rather substantially, as measured by stocking density, kinds and amounts of fertilizer, herbicide use, and length of the culture period. Because of differences in geographic locations, culture activities began earlier in Nebraska than at Valley City, North Dakota; however, at GDNFH, incubation

of walleye was purposely prolonged to delay their hatching until
northern pike were harvested from the ponds that were used for
double-cropping northern pike and walleye. They initially incubated the eggs at 12.8°C for a few days; then the temperature was
lowered to not less than 5.5°C for most of the incubation period
(Zitzow 1991). Differences in mean water temperatures during the
culture interval at the three hatcheries reflected the latitudinal and
management differences. North Platte had the coolest water temperature, and GDNFH had the warmest because GDNFH delayed walleye production. The difference in mean water temperature affected
the length of the culture period: 45 to 59 days at NPSFH (mean temperature ≈ 15°C); 39 to 52 days at VCNFH (mean temperature ≈ 17°C);
but only 24 to 29 days at GDNFH (mean temperature ≈ 20°C). The
culture interval at GDNFH is much smaller than those at the other two
hatcheries in this study; culture intervals used to produce walleye in
other locations are: 31 to 33 days in Illinois (Jahn et al. 1989); 28 to
56 days in Michigan (Gustafson 1991); 44 to 46 days (Fox and
Flowers 1990) and up to 59 days in Ontario (Fox et al. 1989) .

It has often been stated that growth and survival of walleye from
fry to fingerlings in culture ponds is both unpredictable and variable
(Smith and Moyle 1943; Beyerle 1979; Nickum 1986; Qin and
Culver 1992) and that "consistency" across years is one measure of
a successful culture program (Gustafson 1991). Jahn et al. (1989)
reported 55% survival in 1987 and 91% in 1988 at a stocking
density of 250,000 fry/ha. Fox and Flowers (1990) obtained 62.3 to
87.1% survival in eight ponds at stocking densities of 200,000 to
600,000/ha. Qin and Culver (1992) reported that survival ranged
from 77.9% in two ponds fertilized with inorganic fertilizer to only
2.5% in two ponds where oxygen depletion occurred after they
were fertilized with both inorganic and organic fertilizers.

CV should be an appropriate statistic for characterizing consistency among ponds and years. However, to know whether a particular CV was unusually large requires experience with similar data
(Steel and Torrie 1980), which do not exist for walleye fingerling
production. A 19-year data set from NPSFH (1973-1991) provided
information on annual and on pond-to-pond variability. In that data
set, there were greater year-to-year variations in yield and survival
than pond-to-pond variation within years, which suggested that

annual variation in climatic factors (e.g., temperature or incidence of cloudy days) influenced the production process. At NPSFH, the CV for annual yield was 38.7%, compared with an average CV of 30.6% for pond-to-pond variation within years. Likewise, the CV for annual variation in survival was 38.7%, compared with 30.7% for pond-to-pond variation. In this data set, there was a strong linear correlation ($r = 0.83$, $P < 0.01$) between yield and survival. Factors that are likely to affect survival include cannibalism, starvation, the interaction between starvation and cannibalism, occurrence of oxygen depletion, and vegetation problems that hinder harvest (Cuff 1977). Other investigators have noted an inverse relationship between survival and size of walleyes harvested from ponds (Dobie 1956, 1969; Cheshire and Steele 1972).

Correlations between length of the culture period (days) and several production variables were informative. A strong significant negative correlation occurred between number of days in the culture interval (range was 10 days) and survival at NPSFH in 1991; survival decreased an average of 2.3%/day (23% over the 10-day harvest). Although the correlation between survival and culture interval was not significant at NPSFH in 1990 or at the other hatcheries, the finding at NPSFH suggests that if the trophic level on which fry feed was depleted, starvation, cannibalism, or both may rapidly reduce survival. The findings also suggest that efforts to increase average size of the fingerling–without thinning the population with a partial harvest or special efforts to increase food supplies–by extending the culture interval should be undertaken with caution. There was a significant ($P < 0.05$) correlation between length of the culture interval and fish length in 3 of 6 comparisons and for weight in 4 of 6 comparisons.

Fertilizing walleye ponds to increase fish production has been a walleye pond management tool for many years (Smith and Moyle 1943; Dobie 1956), and most fish hatcheries that raise planktivorous fish use organic fertilizers to stimulate production of zooplankton (Geiger 1983; Richard and Hynes 1986; Fox et al. 1989; Jahn et al. 1989). Fertilization was used at the hatcheries in North Dakota, but traditionally, walleye ponds at NPSFH have not been fertilized. At the NPSFH, however, ponds were stocked at 250,000 fry/ha, as compared to 375,000/ha at VCNFH and 394,000/ha at GDNFH.

Higher stocking densities with organic fertilization have been successful at NPSFH (Harding and Summerfelt 1993). Although the focus of pond fertilization has been to develop zooplankton populations, there is no evidence to indicate when mortality takes place in phase I walleye culture. Is the variation in survival due to an inadequate zooplankton community of the right size and kinds for the first feeding larvae, or is poor survival the result of starvation in the last weeks before harvest?

Many different types of and application rates of organic fertilizers have been reported; however, hatchery managers usually have based their fertilizer choices on local availability, cost, and personal experience. Experimental studies using organic fertilizers, however, are available for reference. The comparisons made by Barkoh and Rabeni (1990) on alfalfa meal, wheat shorts, and cottonseed meal for their nutritional value to zooplankton supports the traditional use of alfalfa fertilizers at the federal hatcheries in North Dakota. Jahn et al. (1989) found that organic fertilizers (soybean meal, alfalfa pellets, and torula yeast) were superior to inorganic fertilizer (liquid formula of 10-35-0, N:P:K) in nearly every category of production in plastic-lined ponds; also, the inorganic fertilizers were said to promote poor water quality (high pH, low dissolved oxygen, and high ammonia). However, Qin and Culver (1992) reported that walleye fingerling survival averaged 77.9% in ponds fertilized with inorganic fertilizer alone but that survival averaged only 2.5% in ponds that had been fertilized with both inorganic and organic fertilizers; ponds fertilized with both inorganic and organic fertilizers had an oxygen depletion.

Pond fertilization rates vary considerably. In the present study, ponds at NPSFH were not fertilized, ponds at VCNFH were fertilized with 800 kg/ha of alfalfa pellets, and ponds at GDNFH were fertilized with 1,600 kg/ha of ground alfalfa hay. Richard and Hynes (1986) recommended using "45 kg soymeal/acre-ft/week" (i.e., about 364.8 kg/ha/week) for 6-8 weeks (2,188 to 2,918 kg/ha). Jahn et al. (1989) used 224.2 kg/ha each of alfalfa pellets and soybean meal and 22.4 kg/ha of torula yeast in plastic-lined ponds used to culture walleye. Geiger (1983) obtained 23% survival of phase I striped bass, *Morone saxatilis*, in ponds with 3,080 kg/ha ground Bermuda hay, but obtained 62% survival from ponds fertilized with

2,986 kg/ha of chicken manure. Fox et al. (1989) compared ponds fertilized with fermented soybean meal at 36 g/m^3/week (360 kg/ha) with ponds where the fertilizer was progressively reduced from 32 g/m^3/week until fertilization ceased six weeks after the walleye were stocked. Fish survival, mean length, and yield was higher in ponds given the constant fertilization treatment than in ponds given the progressively reduced fertilization treatment.

A cover crop of rye-grass was planted in the ponds at GDNFH after the ponds were drained. The other two hatcheries did not use this technique. Rye grass prevents erosion of the pond dikes and provides a source of organic fertilizer when the ponds are flooded in the spring. Rye grass may encourage greater chironomid densities in the ponds by serving as an egg-laying platform for the adults when they emerge (Merritt and Cummins 1978). Although most of the nutrients were lost when the ponds were drained and refilled before walleye were stocked, the ponds at GDNFH had nearly four times the chironomid density that was observed at VCNFH.

Fox and Flowers (1990) reported that growth of juvenile walleye was strongly density-dependent, but that survival and density were not related. They suggested that if the production goal is a larger sized fingerling, then a lower yield is required. Differences in fish density are responsible for pond-to-pond variability in growth, because prey densities limit growth but not survival (Fox 1989). However, over the range of densities observed at the three hatcheries in the present study–250,000/ha at NPSFH, 375,000/ha at VCNFH, and 390,000/ha at GDNFH–the correlation between growth rate (mm/day) and density ($r = -0.25$) was not significant ($P = 0.84$). The "standard" stocking density for walleye used at the White Lake facility in Ontario is 400,000/ha (Fox and Flowers 1990). Geiger (1983) stocked striped bass fry at 247,000/ha.

Fox and Flowers (1990) suggested a stocking density of 60 fry/m^3 (600,000/ha for ponds 1 m deep) to maximize juvenile biomass at harvest, an extraordinary recommendation based on the observations and stocking densities reported in this and other studies. The success of a practical stocking density will be a function of ability to provide sufficient forage (especially chironomids) to promote the growth rate needed to achieve desired size at harvest, as well as the desired number of fingerlings at harvest. Cannibalistic behavior of walleye also may play a role in determining higher stocking densities (Dobie 1956; Li and Mathias 1982; Swanson and Ward 1985; McIntyre et al. 1987). Cannibalism of juvenile walleye increased when suitable alternative prey was lacking, and it is also a function of fish density (Li and Mathias 1982; Swanson and Ward 1985).

Herbicide use was variable among these hatcheries, but herbicides were used at all three hatcheries to control *Hydrodicton, Chara, Potamogeton,* or other problem weeds. Herbicides were used mainly to avoid entrapment of the fingerlings during pond draining. The ponds used at GDNFH did not develop nuisance algal problems, but copper sulfate was used on two ponds in 1989 two days prior to harvest. Herbicide applications increase production costs, and they reduce dissolved oxygen and zooplankton densities. However, if herbicide applications are used near the end of production when fingerlings have converted mainly to a chironomid/diet (Fox 1989), the effects of the herbicides on zooplankton may not be of great importance.

Other problems have developed at these hatcheries during the culture period. Clam shrimp, *Cyzicus setosa,* were a problem in 1989 at GDNFH during harvest when they clogged the catch basin screens. Extra cleaning of the screens was required to prevent overflow and loss of fish. At GDNFH, an occasional carryover of northern pike fingerlings which survived in puddles of the drained ponds has reduced survival in the double-cropped ponds. To prevent carryover of northern pike fingerlings, ponds must be dried completely or chemically treated before they are refilled and stocked with walleye fry for a second crop (Zitzow 1991).

Some practices used here, such as fertilization and weed control, may also be used in undrainable ponds, but there are two major differences related to harvest. In undrainable ponds, not all fish are harvested at the end of the culture season; therefore, these ponds require an annual fall chemical treatment (rotenone) to remove residual fish. Harvest methods also differ; in undrainable ponds, fingerlings are harvested with fyke nets accompanied by a copper sulfate treatment (0.15-0.30 ppm) to stimulate fish movement (Gustafson 1991). There are few research reports on management of undrainable ponds, although they are the major culture system used by private walleye producers in Minnesota.

The goals of state and federal agencies for hatchery production of walleye for maintenance stockings need clarification in light of recent stocking evaluation studies (Ellison and Franzin 1992). Fish size and physiological state of the fish affect survival after stocking (Mitzner 1992). Obtaining a larger sized fish with a higher condition factor may be a more important goal than to maximize the number of fingerlings harvested, but findings in this study suggest that a larger size cannot be obtained by simply lengthening the culture season; it will require an increase in chironomids, the major food of walleye near the end of the phase I culture interval, or a reduction in initial stocking density. A standard fertilization regimen–as to kinds, quantities, and application rates–is probably not practical, but given the diversity of fertilizers and application rates reported in the literature, further research is recommended before general guidelines can be offered. The value of a fall crop of rye grass for walleye culture ponds should be evaluated because of its potential benefits to enhance chironomid populations. In hatcheries using surface water supplies, attention should be given to the relationship between the composition of the zooplankton inoculum in the inflowing water and the zooplankton community that develops in the ponds after filling.

ACKNOWLEDGMENTS

We thank the hatchery managers and their staffs at the three hatcheries–Tom Ellis at North Platte State Fish Hatchery, Tom Pruitt at Garrison Dam National Fish Hatchery, and Matt Bernard at Valley City National Fish Hatchery–for providing ponds, facilities, equipment, and hatchery records for this study. We also thank the Nebraska Game and Parks Commission for providing funding for the research done by Lloyd Harding at the North Platt Fish Hatchery, North Platte, Nebraska. Thanks are offered to Richard O. Anderson for his leadership in organizing the symposium of which this paper is a part. The studies at the Garrison Dam and Valley City National Fish hatcheries were carried out with funds provided by the North Central Regional Aquaculture Center Program under grant number 89-38500-4319 from the U.S. Department of Agriculture. Studies at both sites were coordinated and supported by the

Iowa Agriculture and Home Economics Experiment Station, Ames, Iowa, Experiment Station Project No. 2982. This is journal paper number J-14657 of the Iowa Agriculture and Home Economics Experiment Station, Ames, Iowa, Experiment Station Project No. 2982.

REFERENCES

Barkoh, A., and C. F. Rabeni. 1990. Biodegradability and nutritional value to zooplankton of selected organic fertilizers. Progressive Fish-Culturist 52: 19-25.

Barrows, F. T., W. A. Lellis, and J. G. Nickum. 1988. Intensive culture of larval walleyes with dry or formulated feed: Note on swim bladder inflation. Progressive Fish-Culturist 50:160-166.

Beyerle, G. B. 1979. Extensive Culture of Walleye fry in Ponds at the Wolf Lake Hatchery, 1975-1978. Michigan Department of Natural Resources Research Report 1874, Lansing, Michigan.

Cheshire, W.F., and K.L. Steele. 1972. Hatchery rearing of walleyes using artificial food. Progressive Fish-Culturist 34:96-99.

Colesante, R. T., N. B. Youmans, and B. Ziolkoski. 1986. Intensive culture of walleye fry with live food and formulated diets. Progressive Fish-Culturist 48:33-37.

Conover, M. C. 1986. Stocking coolwater species to meet management needs. Pages 31-39 *in* R. H. Stroud, ed. Fish Culture in Fisheries Management. American Fisheries Society, Bethesda, Maryland.

Cuff, W.R. 1977. Initiation and control of cannibalism in larval walleyes. Progressive Fish-Culturist 39:29-32.

Dobie, J. 1956. Walleye pond management in Minnesota. Progressive Fish-Culturist 17:51-57.

Dobie, J. 1969. Growth of walleye and sucker fingerlings in Minnesota rearing ponds. Internationale Vereinigung für Theoretische und Angewandte Limnologie Verhandlungen 17:641-649.

Ellison, D. G., and W. G. Franzin. 1992. Overview of the symposium on walleye stocks and stocking. North American Journal of Fisheries Management 12:271-275.

Fox, M. G. 1989. Effect of prey density and prey size on growth and survival of juvenile walleye (*Stizostedion vitreum vitreum*). Canadian Journal of Fisheries and Aquatic Sciences 46:1323-1328.

Fox, M. G., and D. D. Flowers. 1990. The effect of fish density on growth, survival and production of juvenile walleye in rearing ponds. Transactions of the American Fisheries Society 119:112-121.

Fox, M. G., J. A. Keast, and R. J. Swainson. 1989. The effect of fertilization regime on juvenile walleye growth and prey utilization in rearing ponds. Environmental Biology of Fishes 26:129-142.

FWS (Fish and Wildlife Service). 1991. Fish and Fish Egg Distribution Report of the National Fish Hatchery System. Fiscal Year 1990/Report No. 25. U.S. Department of the Interior, Fish and Wildlife Service, Washington, D.C.

Geiger, J.G. 1983. A review of pond zooplankton production and fertilization for the culture of larval and fingerling striped bass. Aquaculture 35:353-369.

Gustafson, G. 1991. Pond rearing walleyes in Michigan: Fry to advanced fingerling. Pages 262-265 *in* Proceedings of the North Central Aquaculture Conference, Kalamazoo, Michigan, March 18-21, 1991. North Central Regional Aquaculture Center, Michigan State University, East Lansing, Michigan.

Harding, L. M., and R. C. Summerfelt. 1993. Effects of fertilization and of fry stocking density on pond production of fingerling walleye, *Stizostedion vitreum*. Journal of Applied Aquaculture 2 (3/4):59-79.

Heidinger, R. C., J. H. Waddell, and B. L. Tetzlaff. 1985. Relative survival of walleye fry versus fingerlings in two Illinois reservoirs. Proceedings of the Southeastern Association of Fish and Wildlife Agencies 39:306-312.

Jahn, L. A., L. M. O'Flaherty, G.M. Quartucci, J. H. Kim, and X. Mao. 1989. Analysis of Culture Ponds to Enhance Fingerling Production. Completion Report, Federal Aid Project F-50-R, Illinois Department of Conservation, Springfield, Illinois.

Kendall, R. L. 1978. Selected Coolwater Fishes of North America. American Fisheries Society, Washington, D.C.

Kuss, S.M. 1988. Effect of Fertilization Rate and Zooplankton Inoculation on Zooplankton Populations in Earthen Ponds Without Fish. Master's thesis. State University of New York College at Brockport, Brockport, New York.

Li, S., and J. A. Mathias. 1982. Causes of high mortality among cultured larval walleyes. Transactions of the American Fisheries Society 111:710-721.

McIntyre, D.B., F.J. Ward, and G.M. Swanson. 1987. Factors affecting cannibalism by pond-reared juvenile walleyes. Progressive Fish-Culturist 49:264-269.

Mathias, J.A., and S. Li. 1982. Feeding habits of walleye larvae and juveniles: Comparative laboratory and field studies. Transactions of the American Fisheries Society 111:722-735.

Merna, J.W. 1977. Food Selection by Walleye Fry. Michigan Department of Natural Resources, Fish Research Report No. 1847, Lansing, Michigan.

Merritt, R. W., and K. W. Cummins. 1978. An Introduction to the Aquatic Insects of North America. Kendall /Hunt Publishing, Dubuque, Iowa.

Mitzner, L. 1992. Evaluation of walleye fingerling and fry stocking in Rathbun Lake, Iowa. North American Journal of Fisheries Management 12:321-328.

Nickum, J.G. 1978. Intensive culture of walleyes: The state of the art. American Fisheries Society Special Publication 11:187-194.

Nickum, J. G. 1986. Walleye. Pages 115-126 *in* R. R. Stickney, ed., Culture of Nonsalmonid Freshwater Fishes. CRC Press, Inc., Boca Raton, Florida.

Paragamian, V. L., and R. Kingery. 1992. A comparison of walleye fry and fingerling stockings in three rivers in Iowa. North American Journal of Fisheries Management 12:313-320.

Piper, R. G., I. B. McElwain, L. E. Orme, J. P. McCraren, L. G. Fowler, and J. R.

Leonard. 1982. Fish Hatchery Management. United States Fish and Wildlife Service, Washington, D.C.

Qin, J., and D. A. Culver. 1992. The survival and growth of larval walleye, *Stizostedion vitreum,* and trophic dynamics in fertilized ponds. Aquaculture 108:247-276.

Raisanen, G. A., and R. L. Applegate. 1983. Prey selection of walleye fry in an experimental system. Progressive Fish-Culturist 45:209-214.

Richard, P. D., and J. Hynes. 1986. Walleye Culture Manual. Ontario Ministry of Natural Resources, Fish Culture Section, Toronto, Ontario.

Smith, L. L., and J. B. Moyle. 1943. Factors influencing production of yellow pike-perch, (*Stizostedion vitreum vitreum*) in Minnesota rearing ponds. Transactions of the American Fisheries Society 73:243-261.

Steel, R. G. D., and J. H. Torrie. 1980. Principles and Procedures of Statistics, 2nd ed. McGraw-Hill Book Co., New York, New York.

Swanson, G. M., and F. J. Ward. 1985. Growth of juvenile walleye, (*Stizostedion vitreum vitreum*), in two man-made ponds in Winnipeg, Canada. Internationale Vereinigung für Angewandte Limnologie Verhandlungen 22:2502-2507.

Zitzow, R. E. 1991. Extended incubation of walleye eggs with low-flow incubators. Progressive Fish-Culturist 53:188-189.

Effects of Fertilization and of Fry Stocking Density on Pond Production of Fingerling Walleye, *Stizostedion vitreum*

Lloyd M. Harding
Robert C. Summerfelt

ABSTRACT. The influence of fertilization and of fry stocking density on production of fingering walleye, *Stizostedion vitreum,* was evaluated in earthen ponds at North Platte State Fish Hatchery, North Platte, Nebraska. In 1990, five 0.4-ha ponds were fertilized with alfalfa pellets, and five were fertilized with soybean meal; four unfertilized ponds served as controls. All ponds were stocked with D2 (D1 = the day at hatch) walleye fry at 250,000/ha. Differences in yield, number of fingerlings harvested, mean length, and mean weight among treatments were not statistically significant ($P > 0.05$). In 1991, two fertilization schedules (no fertilizer and fertilization with alfalfa pellets) and two fry stocking rates (250,000 and 375,000 fry/ha) were evaluated. Four ponds were used for each treatment. Statistically significant treatment differences were found in yield,

Lloyd M. Harding and Robert C. Summerfelt, Department of Animal Ecology, Iowa State University, Ames, IA 50011-3221, USA.

Correspondence for Lloyd M. Harding may be addressed to Coldwater Fish Farms, Inc., P.O. Box 1, Lisco, Nebraska 69148, USA.

[Haworth co-indexing entry note]: "Effects of Fertilization and of Fry Stocking Density on Pond Production of Fingerling Walleye, *Stizostedion vitreum.*" Harding, Lloyd M., and Robert C. Summerfelt. Co-published simultaneously in the *Journal of Applied Aquaculture,* (The Haworth Press, Inc.) Vol. 2, No. 3/4, 1993, pp. 59-79; and: *Strategies and Tactics for Management of Fertilized Hatchery Ponds* (ed: Richard O. Anderson and Douglas Tave) The Haworth Press, Inc., 1993, pp. 59-79. Multiple copies of this article/chapter may be purchased from The Haworth Document Delivery Center [1-800-3-HA-WORTH; 9:00 a.m. - 5:00 p.m. (EST)].

59

number of fingerlings harvested/ha, average length, and average weight. Yield was higher in fertilized ponds compared with yield from unfertilized ponds at both stocking densities, but yield did not differ significantly between stocking density treatments given the same fertilizer treatment. Survival did not differ between density treatments, but total number of fish harvested was significantly greater from ponds stocked at the higher density. Fingerlings with the largest average weight were raised in fertilized ponds that were stocked at 250,000/ha, while the smallest fingerlings were from unfertilized ponds that were stocked at 375,000/ha. Days in culture interval, which varied among ponds by 9 days in 1990 and 10 days in 1991, was significantly correlated with most production variables in 1990 and with all production variables in 1991. Means of water quality variables were not significantly different between fertilized and unfertilized ponds in either year, but significant differences were found in means of three water quality variables between 1990 and 1991. Yield in both fertilized and unfertilized ponds in 1991 was less than in 1990.

INTRODUCTION

Pond-raised walleye, *Stizostedion vitreum*, fingerlings are widely used in the United States and Canada for maintenance stockings (Li and Ayles 1981; Conover 1986; Ellison and Franzin 1992), but demand for fingerling walleye often exceeds available supply (Buttner 1989). Walleye fingerlings are raised in undrainable ponds, drainable ponds, and plastic-lined ponds; water supplies include surface and ground water. Because of this variability, many cultural practices are site-specific, and no universal management techniques have developed (Mathias and Li 1982; Richard and Hynes 1986; Buttner 1989). Most cultural practices for walleye have developed from trial-and-error rather than from experimental design (Richard and Hynes 1986). Even in well-managed, drainable ponds (i.e., with fertilizers to enhance production and herbicides for weed control), both annual and pond-to-pond variation in survival and yield of fingerlings from ponds has been quite variable (Smith and Moyle 1943; Cheshire and Steele 1972; Qin and Culver 1992).

The range in stocking density at production hatcheries and in experimental studies varied more than 20-fold (29,000 to 600,000/ ha; Table 1). To realize both optimal utilization of available pond

TABLE 1. Walleye stocking densities in earthen ponds.

Stocking density (No./ha)[a]	Reference
600,000	Fox and Flowers (1990)
437,500	Fox et al. (1989)
400,000	Fox (1989)
400,000	Fox and Flowers (1990)
394,000	Clouse (1991)
375,000	Richard and Hynes (1986)
250,000	Jahn et al. (1989)
244,000	Qin and Culver (1992)
200,000	Fox and Flowers (1990)
75,000	Buttner et al. (1991)
50,000	Buttner (1989)
38,000	Swanson and Ward (1985)
29,000	Swanson and Ward (1985)

[a]When stocking density was reported as No./m^3, the No./ha was estimated by assuming a mean pond depth of 1 m.

resources and production potential of the ponds, high stocking densities are needed, but an inverse relation has been reported for size of walleye harvested from ponds and their rate of survival (Dobie 1956; Cheshire and Steele 1972). Variables that have determined fry stocking densities include water quality, fertility, and size and number of fingerlings desired (Rees and Cook 1983; Fox and Flowers 1990).

Fertilized ponds may be stocked at higher densities, but density-dependent mortality may limit survival. Li and Mathias (1982) showed density-dependent survival of postlarval (10-19 mm total length) walleye in aquarium studies; however, in small ponds, Fox and Flowers (1990) reported density-independent survival at fry stocking rates between 200,000 to 600,000 fry/ha. In lakes, the survival rates of larvae and juvenile walleye seem to be density-independent (Forney 1976; Carlander and Payne 1977), but densities in lakes are far lower than those in ponds used for walleye culture.

Dobie (1956) was the first to note a relationship between pond substrate fertility and walleye production. Pond fertilization has resulted in increased fingerling yield and improved survival (Smith and Moyle 1943; Dobie 1956; Geiger et al. 1985; Buttner 1989; Fox et al. 1989). However, optimal fertilizer types, rates, and application regimens have yet to be identified for walleye culture. Inorganic fertilizers have not been commonly used in walleye culture (Richard and Hynes 1986). Jahn et al. (1989) recommended using only organic fertilizers for walleye fingerling production in plastic-lined ponds because inorganic fertilizers promoted high pH, low dissolved oxygen, and high ammonia.

Organic fertilizers may stimulate algae through a slow release of inorganic macronutrients from decomposition, and the organic matter (organic carbon) may directly feed heterotrophic pathways (Knud-Hansen et al. 1991). An abundance and diversity of organisms thrive on organic matter and rapidly convert it to high quality fish food organisms (Schroeder 1978). Organic debris and microbes utilizing it constitute an important part of the diet of zooplankton (Makarewicz and Likens 1979; Glamazda and Katretskiy 1980), which in turn are an important part of the diet of fingerling fish. Chironomid larvae also use organic matter in the pond substrate (Oliver 1971); therefore, organic fertilizers may be especially important during the second half of the culture period, when walleye shift to a diet dominated by chironomids (Fox et al. 1989; Jahn et al. 1989; Clouse 1991; Harding 1991).

Walleye culturists have used mostly forms of alfalfa (chopped hay, meal, or pellets), soybean meal (Richard and Hynes 1986; Jahn et al. 1989; Qin and Culver 1992; Summerfelt et al. 1993) or fermented soybean meal (Richard and Hynes 1986; Fox and Flowers 1990). Geiger et al. (1985) recommended that the application rates of organic fertilizer be based on nitrogen content; i.e., different types of organic matter should be applied on the basis of similar total nitrogen.

The objectives of the present study were to evaluate fertilization and stocking density strategies for improving production of phase I walleye in earthen ponds. Two types of organic fertilizers–soybean meal and alfalfa pellet–were applied on the basis of similar total nitrogen but different total biomass, and two fry stocking rates

(250,000 and 375,000 fry/ha) were compared in fertilized and un-fertilized ponds.

MATERIALS AND METHODS

Study Site

Experiments were conducted in 0.4-ha earthen ponds (mean depth about 1 m) at the Nebraska Game and Parks Commission's North Platte State Fish Hatchery (NPSFH), North Platte, Nebraska. The ponds were filled with water drawn from the Nebraska Public Power District's hydro-electric/irrigation canal. The canal water originates from Lake McConaughy, an impoundment on the North Platte River; before reaching the hatchery, the water flows through several irrigation reservoirs in west-central Nebraska, the last of which is Lake Maloney, located a short distance above the point at which the NPSFH draws its water. The water supply had a total alkalinity of 150-200 mg/L as $CaCO_3$ and a pH generally between 8.0 and 9.0.

Pond Filling

The ponds at NPSFH required 1 to 3 days to fill, but all ponds were filled the same week. Filling began approximately 10 days before stocking. After the ponds were filled, the inflow was reduced to a rate adequate to compensate for water lost to either seepage or to evaporation. The amount of water added (300 L/minute) was equivalent to a complete turnover of water volume in each pond every 10 days during the culture period.

Stocking

D2 walleye fry were stocked April 13-16, 1990 and April 14-15, 1991. When fry were stocked, water temperatures were 7.5°C in 1990 and 10°C in 1991. Fry were stocked in the mornings on the leeward side of each pond. Counts of fry for stocking were estimated by volumetric methods (Piper et al. 1982). In 1990, all ponds were stocked at 250,000/ha, the rate traditionally used at this hatchery for phase I walleye fingerling production. In 1991, two stocking densities were evaluated: 250,000 fry/ha and 375,000 fry/ha.

Vegetation Control

Rooted aquatic macrophytes were treated with simazine (Aquazine™;[1] 90% active ingredient) during both years. Simazine was applied to all ponds with a broadcast sprayer, at a rate of 24.2 kg/ha (2.2 ppm) during the fourth week of culture in 1990 and during the fifth week in 1991.

Water Quality

Selected water quality variables were monitored weekly each year. Water temperature and dissolved oxygen were measured with a dissolved oxygen meter and temperature with a thermistor. In 1990, dissolved oxygen readings were taken at a depth of 0.5 m at sundown and three hours later to predict a pre-sunrise concentration. In 1991, pre-sunrise oxygen measurements were taken.

Total ammonia-nitrogen (TAN) was determined by the Nesslerization method; nitrate-nitrogen was determined by the cadmium reduction method; and nitrite-nitrogen was determined by the colorimetric sulfanilamide method (APHA et al. 1989). The pH of pond waters was measured at the pond surface near mid-day (1100-1400 hours) by means of a pH meter fitted with a temperature compensating glass electrode. Un-ionized ammonia (UIA) was obtained by multiplication of percent un-ionized ammonia at specific temperature and pH values from tables given by Thurston et al. (1979). Secchi disk depth was measured near mid-day (1100-1400 hours) at the deepest part of the ponds.

Fertilization

In 1990, two types of organic fertilizers (soybean meal and alfalfa pellets) were each applied at 23.0 kg N/ha over 6 weeks. Nitrogen

---

1. Use of trade or manufacturer names does not imply endorsement.

content was calculated from the crude protein concentration based on the feed label; i.e., N = (protein/100) × 16 (NRC 1982). Subsequent proximate analysis, however, indicated that less nitrogen was applied to the ponds receiving alfalfa pellets (20.7 kg N/ha) than to those receiving soybean meal (23.1 kg N/ha). Both alfalfa pellets and soybean meal were used in 1990, because at the same nitrogen application rate they differed substantially in biomass. Each fertilizer treatment was applied to each of five ponds; four ponds were not fertilized and served as the control. Fertilizer was applied four times; applications began at pond filling and they were weekly for the next 3 weeks. Fertilizer was distributed evenly over the ponds; each application was either 212 kg/ha of alfalfa pellets or 88 kg/ha of soybean meal. Total applications were 848 kg/ha of alfalfa pellets or 352 kg/ha of soybean meal.

In 1991, only alfalfa pellets were used as fertilizer. Alfalfa pellets were applied to eight ponds; eight other ponds served as the no fertilizer control. Fertilizer was applied six times; applications commenced at pond filling, and they were weekly thereafter at 141 kg/ha applications; total amount of fertilizer applied was 848 kg/ha.

Zooplankton Sampling

Zooplankton densities in the ponds and in the inflow water were sampled weekly. Samples in the ponds were collected at night by means of a flexible impeller pump and a Wisconsin plankton net (80 μm mesh), through which inflow samples were also filtered. Plankton samples were fixed and stored in 4% buffered formalin solution with sucrose until analyzed. Subsamples were removed from the original samples with a 1-ml Hensen-Stempel pipette, then enumerated and identified in a plankton counting wheel mounted on a dissecting microscope, and viewed at a magnification of 4x. No less than 100 organisms were counted in each subsample. Three subsamples of the original sample were counted and averaged; the number/L for each sample was calculated from the mean of the sub-samples and application factors for the sample dilutions.

Fish Harvest

Forty-five to 59 days after stocking (Table 2), water was drained from the ponds into outside catch basins where the fingerlings were

TABLE 2. The number of culture days and the correlation (r) between culture days and production statistics for the 1990 and 1991 experiments. Correlations followed by one asterisk are significant at $P < 0.05$; those followed by two asterisks are significant at $P < 0.01$.

Variable	1990 experiments	1991 experiments
Culture days		
mean	50.3	53.7
range	45-54	49-59
	Correlations (r)	Correlations (r)
Survival (%)	—	-0.77**
Yield (kg/ha)	0.77**	0.85**
No/ha	-0.42	-0.90**
Mean length	0.84**	0.89**
Mean weight	0.67*	0.96**

harvested. Fingerling biomass at harvest was determined by weighing all fish. Three to four sample counts were used to estimate the number of fingerlings/kg and the number of fingerlings harvested from each pond was estimated. Survival was determined by dividing the estimated number of fingerlings harvested by the estimated number of fry stocked. At harvest each year, 20 fingerlings were collected and preserved in 10% buffered formalin for analysis of diet. An additional 20 to 25 fish were measured live to determine the total length at harvest.

Statistical Analysis

Differences in production data from 1990 were assessed by a one-way analysis of variance (SAS Institute, Inc., 1985). Four ponds were used for the control treatment and five ponds for each of the two fertilizer treatments (soybean meal and alfalfa pellets). The 1991 experiment was a 2 × 2 factorial design in which there were two levels of fertilizer and two stocking densities. The four combinations were each replicated four times (a total of 16 ponds). The fertilizer factor treatments were no fertilization and fertilization with alfalfa pellets; the stocking density factor treatments were 250,000 and 375,000/ha. The traditional treatment at NPSFH had been to use no fertilizer and to stock at 250,000 fry/ha. A two-way analysis of variance procedure was used to assess differences. The length of the culture interval varied 9 to 10 days between ponds, because all ponds were not stocked or harvested on the same dates; therefore, survival, number harvested/ha, mean length, and mean weight were adjusted by analysis of covariance (Steel and Torrie 1980), using the number of days in the culture interval as the covariate. Differences between treatments or among treatment combinations were considered to be significant at $P < 0.05$. Differences in means of water quality characteristics between treatments were assessed using one-way analysis of variance, with three treatments in 1990 and four in 1991.

Because the number harvested/ha and the distribution of proportions may not be normally distributed, all other production variables (number harvested/ha, survival, yield, mean length, and mean weight) were checked for normality using the univariate procedure (SAS Institute, Inc. 1985). None of the distributions showed a significant departure from normality; therefore, the data were used without transformation.

RESULTS

Water Quality

In both 1990 and 1991, the differences among the means of water quality measurements between fertilized and unfertilized ponds were remarkably small; there were no significant differences among the experimental treatments (Table 3). The highest weekly mean concentration of un-ionized ammonia was 0.092 mg/L, which is greater than the 0.0125 mg/L ammonia criteria recommended for optimum health of salmonids (Piper et al. 1982), but it was far less than 96-hr LC_{50} values (Boyd 1990). Pre-sunrise oxygen concentrations were never threatening (minimal value was 4.2 mg/L) to survival, and emergency aeration was not required.

Water quality differences between 1990 and 1991 were greater than treatment differences within those years (Table 4). Unfertilized ponds that were stocked with 250,000 fry/ha in both 1990 and 1991

TABLE 3. Treatment means of water quality variables in walleye culture ponds in 1990 and 1991. In 1991, each treatment was a combination of two factors: fertilizer (no fert = no fertilizer, alfalfa = fertilized with alfalfa pellets) and stocking density (low = 250,000 fry/ha, and high = 375,000 fry/ha).

Treatment	Temp. (°C)	pH	TAN[a] (mg/l)	UIA[b] (mg/l)	Nitrite (mg/l)	Nitrate (m/l)	Secchi disk depth (cm)	D.O.[c] (mg/l)
				1990				
Alfalfa pellets	16.4	8.8	0.566	0.087	0.021	0.522	85.2	7.3
Soybean meal	16.5	8.8	0.578	0.092	0.019	0.450	78.1	7.4
No fertilizer	16.5	8.9	0.530	0.091	0.019	0.447	82.4	7.7
F-value for ANOVA	0.002	0.096	0.045	0.033	0.432	0.125	0.081	0.027
P values	0.992	0.909	0.959	0.968	0.657	0.884	0.923	0.974
				1991				
No fert x low	14.2	8.5	0.138	0.010	0.013	0.467	97.1	9.1
No fert x high	14.1	8.5	0.116	0.009	0.012	0.467	94.6	9.0
Alfalfa x low	14.1	8.4	0.127	0.009	0.012	0.433	82.5	9.0
Alfalfa x high	14.3	8.4	0.134	0.009	0.013	0.450	85.0	8.9
F-value for ANOVA	0.001	0.119	0.078	0.113	0.648	0.014	0.282	0.006
P values	0.999	0.947	0.970	0.950	0.606	0.997	0.837	0.999

[a]Total ammonia nitrogen
[b]Un-ionized ammonia calculated from tables given by Thurston et al. (1979)
[c]Dissolved oxygen in 1990 was the projected pre-sunrise concentration (Dupree and Huner 1984); the 1991 data represent measured pre-sunrise concentration.

had significantly ($P < 0.05$) different concentrations of TAN and UIA. Ponds that were fertilized with alfalfa pellets and that were stocked with 250,000 fry/ha had significantly different ($P < 0.05$) concentrations of TAN, UIA, and nitrite in 1990 and 1991 (Table 4).

Zooplankton

Except for the first week of culture week in 1990, the numbers of copepods and cladocerans in the ponds during both years were always greater than 200 zooplankton/L (Figure 1). Other zooplank-

TABLE 4. Comparison (one-way analysis of variance) of year-to-year (1990 and 1991) differences in means of water quality variables in unfertilized and fertilized ponds that were stocked 250,000 fry/ha.

Year	Temp. (°C)	pH	TAN (mg/l)	UIA (mg/l)	Nitrite (mg/l)	Nitrate (m/l)	Secchi disk depth (cm)	D.O.[a] (mg/l)
Unfertilized ponds stocked with 250,000 fry/ha								
1990	16.5	8.9	0.530	0.091	0.019	0.447	82.4	7.7
1991	14.2	8.5	0.138	0.010	0.013	0.446	97.1	9.1
F-value for ANOVA	0.600	1.799	5.661	14.578	3.646	0.519	0.519	0.552
P values	0.464	0.222	0.048	0.007	0.098	0.4994	0.494	0.482
Fertilized (alfalfa pellets) ponds stocked with 250,000 fry/ha								
1990	16.4	8.8	0.566	0.087	0.021	0.522	85.2	7.3
1991	14.1	8.4	0.127	0.009	0.012	0.433	82.5	9.0
F-value for ANOVA	0.576	1.078	7.596	10.527	13.235	0.016	0.016	1.056
P values	0.472	0.334	0.028	0.014	0.008	0.904	0.904	0.338

[a]Total ammonia nitrogen
[b]Un-ionized ammonia calculated from tables given by Thurston et al. (1979)
[c]Dissolved oxygen in 1990 was the projected pre-sunrise concentration (Dupree and Huner 1984); the 1991 data represent measured pre-sunrise concentration.

ton, primarily rotifers, made up a substantial portion of the total zooplankton density, especially during the first 3 to 4 weeks of the culture period, but they were never found in walleye gut contents (Harding 1991). In 1990, the combined abundance of copepods and cladoceran peaked at 2,000/L in the fifth culture week. In 1991, the combined copepod and cladoceran density was 1,000/L by the third week. Combined density remained at nearly the same level through the sixth week of culture before declining to 250/L at harvest. Mean combined copepod and cladoceran densities through the culture period were 357/L in 1990 and 544/L in 1991.

Samples of the inflow water throughout the culture period contained an average of 490 zooplankton/L in 1990 and 506 zooplankton/L in 1991. With an average inflow of 300 L/minute and an average zooplankton density of about 500/L, the inflow would add 216 million zooplankton/day. The daily inflow of zooplankton

FIGURE 1. Averages of weekly measurments of copepods, cladocerans, and other zooplankton (mainly rotifers) densities in earthen culture ponds at the North Platte State Fish Hatchery in 1990 and 1991.

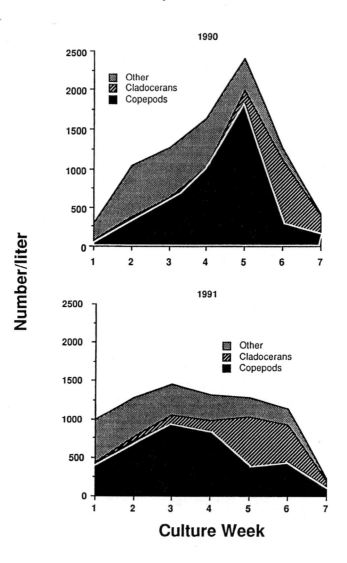

would represent an average density of 54 zooplankton/L in the ponds; therefore, the contribution of zooplankton by the inflow would be about 15% of the average pond density in 1990 and 9.9% in 1991.

Treatment Effects

There were no significant differences in water quality between treatments in 1990 or 1991 (see water quality section, page 67). In 1990, production variables were not significantly different among the three treatments (Table 5). However, in 1991, there were significant differences ($P < 0.05$) among the four treatment combinations for yield, number/ha, mean weight, and mean length at harvest (Table 6).

In 1991, survival rates among the treatments did not differ significantly, but differences between treatments were significant for yield, total number of fingerlings harvested/ha, average length, and average weight at harvest. There were no fertilizer-stocking density interactions affecting yield, number of fingerlings harvested/ha, or fingerling length. The only significant interaction between fertilizer and fry density was for fingerling weight ($P < 0.05$). Except for mean weight, fertilizer and stocking density effects were analyzed separately.

Duncan's multiple comparison procedure indicated that yield in 1991 was greater in ponds fertilized with alfalfa pellets than in

TABLE 5. Comparison of 1990 walleye fingerling production (means±SD) by fertilizer treatment (*P* values for analysis of covariance of treatment effects).

| | Fertilizer Type | | | |
	None (control)	Soybean meal	Alfalfa pellets	*P*
Yield (kg/ha)	87.4±6.7	82.4±5.4	89.4±5.7	0.64
No. harvested/ha	256,523±6,029	243,800±4,819	260,269±5,065	0.61
Mean length (mm)	35.5±0.9	35.9±0.7	37.0±0.7	0.44
Mean weight (g)	0.29±0.03	0.32±0.03	0.35±0.03	0.47
No. of ponds	4	5	5	

TABLE 6. Analysis of fertilizer and stocking density factors for the 1991 walleye fingerling production (treatment means±SD) experiments (*P* values for analysis of covariance of treatment effects). Values in a row followed by the same letter are not significantly different (*P* < 0.05).

| | No fertilizer | | Fertilized with alfalfa pellet | | |
	250,000/ha	375,000/ha	250,000/ha	375,000/ha	P
Survival (%)	61.8±4.9a	61.3±4.9a	67.5±4.9a	68.4±4.9a	0.64
Yield (kg/ha)	52.0±4.7a	57.2±4.7a	77.3±4.7b	81.0±4.7b	0.01
No. harvested/ha	153,509± 5,900a	228,647± 6,043b	167,277±15,871a	253,197±15,871b	0.01
Mean length (mm)	36.2±0.9ab	33.5±0.9c	38.3± 0.9b	35.8 ± 0.9a	0.02
Mean weight (g)	0.33 ± 0.02a	0.28 ± 0.02b	0.47 ± 0.02c	0.32 ± 0.02a	0.01
No. of ponds	4	4	4	4	

unfertilized ponds, but yield did not differ significantly between stocking density treatments for ponds with the same fertilizer treatment (Table 6). The total number of fish harvested did not differ between fertilizer treatments, but more fish were harvested from ponds stocked with 375,000/ha than from ponds stocked with 250,000/ha. The largest mean length was for fish raised in fertilized ponds that were stocked at 250,000/ha; the smallest fish came from ponds without fertilizer that were and stocked at 375,000/ha.

DISCUSSION

In 1990, at a stocking density of 250,000/ha, there were no significant differences in total yield, number of fingerlings harvested/ha, mean length, and mean weight at harvest in fertilized and unfertilized ponds. Equality in yield of fingerlings from the two fertilizer treatments, with about equal nitrogen applications but with large differences in the biomass applied, would support the recommendation by Geiger et al. (1985) that organic fertilizers be applied on the basis of similar total nitrogen. However, the relative merits of that recommendation cannot be confirmed by the findings of the present study, given that yield differences between the fertilized and unfertilized ponds were not significant; clearly, the 250,000 fry/ha stocking rates in 1990 did not exceed the production limit of the ponds derived from basic fertility of the pond soil and inflowing water. Ludwig and Tackett (1991), however, indicated that the rate of nitrogen application may not be the most important consideration in selection of organic fertilizers.

The 1990 results did not provide guidance on which fertilizer to use in the 1991 experiments. The decision to use alfalfa pellets as the organic fertilizer in 1991 was based on other factors: the reported nutritional value of alfalfa to zooplankton (Barkoh and Rabeni 1990), the potential value of alfalfa as a food source to chironomid larvae (Oliver 1971), and the knowledge that alfalfa pellets and ground alfalfa hay have been the traditional fertilizer used in large federal hatcheries in North Dakota (Summerfelt et al. 1993). Jahn et al. (1989) recommended a combination of alfalfa pellets, soybean meal, and torula yeast for fertilizing undrainable walleye culture ponds.

In 1991, yield was significantly higher in the fertilized ponds at both stocking densities than in the unfertilized ponds. Average yield was 45% greater from ponds fertilized with alfalfa pellets (79.2 kg/ha) than the yield from unfertilized ponds (54.6 kg/ha). The interaction between the factors for yield and stocking density were not significant.

The average of the yields in 1990 and 1991 from ponds at the North Platte hatchery that were stocked at 250,000 fry/ha and fertilized with alfalfa pellets was 83.4 kg/ha, and when ponds were stocked with 375,000 fry/ha the highest yield from fertilized ponds was 81.0 kg/ha. These yields are substantially greater than yields in 1989 and 1990 at two federal hatcheries in North Dakota (Summerfelt et al. 1993). The yields at Valley City National Fish Hatchery averaged 45.4 kg/ha from ponds stocked at 375,000/ha and fertilized with 800 kg/ha alfalfa pellets; yields at Garrison Dam National Fish Hatchery averaged 47.9 kg/ha from ponds stocked at 390,000/ha and fertilized with 1,600 kg/ha ground alfalfa hay (Summerfelt et al. 1993). However, yields of phase I walleye fingerlings from experimental studies in small ponds have been much higher than typical production ponds. Jahn et al. (1989) reported yields of 147.7 kg/ha from five 0.2-ha ponds stocked at 250,000/ha. Fox and Flowers (1990) reported yields of 19.7 g/m^3 (197 kg/ha) at a stocking density of 60/m^3 (600,000/ha) from 3 small (0.04-0.15 ha) experimental ponds.

Except during the first culture week in 1990, zooplankton densities were always higher than the minimal level of 100/L suggested by Mathias and Li (1982) for walleye raised in tanks. The contribution of zooplankton by the inflowing water (15% of the average pond density in 1990) was a substantial inoculum, but the major source of zooplankton in the ponds had to be from *in situ* production. Zooplankton inoculation, however, has been suggested as a means of optimizing production of planktivorous larval fish (Geiger et al. 1985; Richard and Hynes 1986).

Comparison of average yields between 1990 and 1991 from unfertilized ponds stocked at 250,000 fry/ha was 40.5% less in 1991 (52.0 kg/ha) than in 1990 (average yield was 87.4 kg/ha in 1990). Also, yield from ponds stocked at 250,000 fry/ha and fertilized with alfalfa pellets was 32% less in 1991 (89.4 kg/ha) than in 1990 (67.5 kg/ha). In examining a 19-year data base on yield from this hatchery, larger year-to-year variation in mean yield than in within-year variation among ponds was common (Summerfelt et al. 1993). Annual means of some water quality variables in 1990 and 1991, which were remarkably similar between treatments within those years, were significantly different. There were significant year-to-year differences in TAN and UIA in unfertilized ponds stocked with 250,000 fry/ha, and there were significant differences in TAN, UIA, and nitrite in ponds fertilized with alfalfa pellets and stocked with 250,000 fry/ha. Temperature differences between years were not significant, but the average difference was 2.3°C, lending support to a hypothesis that cooler, cloudy weather in 1991 reduced autotrophic production enough to account for the chemical differences. Circumstantial evidence supports this hypothesis; while not statistically significant, Secchi disk depth, which would be would be expected to be greater with less algal production, was greater in 1991 than in 1990. Also, pre-sunrise oxygen concentration, while also not statistically significant, was greater in 1991 than in 1990, which suggests that the magnitude of nocturnal community respiration was higher in 1990 than in 1991.

The 250,000 fry/ha traditional stocking rate used at NPSFH is the same used by Jahn et al. (1989) and Qin and Culver (1992). The 600,000 fry/ha rate reported by Fox and Flowers (1990) from three small (0.04-0.15) experimental ponds is the highest stocking density found in the literature (Table 1). Densities greater than 400,000 fry/ha have not been reported from production hatcheries. The 375,000 fry/ha stocking density used at NPSFH in 1991, however, is probably less than maximum, if fertilizer rates are increased or smaller sized fish at harvest is acceptable. The generalization that the number of fish harvested is linearly related to fish stocking density (Schroeder 1978) was supported by the 1991 stocking density comparison–harvest (number./ha) from ponds stocked at 375,000 fry/ha in unfertilized and fertilized ponds was significantly greater than from ponds stocked at 250,000/ha.

Volumetric methods for estimating fry stocking densities are not as accurate as gravimetric or some mechanical counting devices (Kindschi and Barrows 1991). However, they are commonly used by large production hatcheries. In 1990, routine application of the volumetric method used to estimate numbers of fry for stocking resulted in >100% survival from a few ponds; i.e., the number/ha at stocking was underestimated. In 1991, average survival for 16 ponds was 64.7% (range for individual treatments was 61.3 to 68.4%). Comparative survival values are: 62 to 88% (Fox and Flowers 1990); 77.9% (Qin and Culver 1992); 55 to 91% (Jahn et al. 1989). In this study, survival from ponds stocked at 250,000/ha and 375,000/ha did not differ significantly; somewhat higher stocking rates may further increase the number of fingerlings harvested/ha. In both fertilized and unfertilized ponds, mean length and weight of fingerlings was larger from ponds stocked at 250,000/ha than from ponds stocked at 375,000/ha. The results from the present study and that of Fox and Flowers (1990) do not support previous findings of density dependent survival (Li and Mathias 1982; Rees and Cook 1983).

The production process and density-dependent effects in walleye culture have not been resolved to the point where specific stocking densities and fertilizer regimens can be prescribed to produce specific sized fish. These results, however, suggest a density-dependent relationship between fingerling length and weight at harvest. In 1991, more fingerlings were harvested at the higher stocking density, but the mean length and weight at harvest was smaller from ponds stocked at the higher density. There are conflicting reports on the effects of fry density on growth. Dobie (1956) and Fox and

Flowers (1990) reported an inverse relationship between stocking density and size at harvest, whereas Li and Ayles (1981) did not.

Ponds are not filled simultaneously or stocked on the same day, and harvest of 16 ponds takes 9-10 days; thus, the length of the culture interval varies 9 to 10 days among ponds. The number of culture days in the growing season for phase I walleyes at this hatchery had a significant effect on production parameters in both years. The data indicate that trying to increase fingerling size by lengthening the culture period will be offset by a reduction in number of fingerlings harvested if the stocking density is close to maximum carrying capacity of the ponds. However, the findings demonstrate that organic fertilizer application and increased stocking densities under careful management can ensure that desired size and number of fingerlings are produced and can also help reduce year-to-year and pond-to-pond variability.

ACKNOWLEDGMENTS

This is journal paper number J-14673 of the Iowa Agriculture and Home Economics Experiment Station, Ames, Iowa; Project Number 2982. This project is a result of work sponsored by the Nebraska Game and Parks Commission and the Iowa Agricultural and Home Economics Experiment Station. The Fisheries Division of the Nebraska Game and Parks Commission provided the culture ponds and equipment used at the North Platte State Fish Hatchery. We thank Fisheries Division Chief Wes Sheets and Fisheries Production Supervisor Larry Zadina, as well as Hatchery Superintendent Tom Ellis and his staff at North Platte State Fish Hatchery for their cooperation and assistance with this project.

REFERENCES

APHA et al. (American Public Health Association, American Water Works Association, and Water Pollution Control Federation). 1989. Standard Methods for the Examination of Water and Wastewater, 17th ed., American Public Health Association, Washington, D.C.

Barkoh, A., and C. F. Rabeni. 1990. Biodegradability and nutritional value to zooplankton of selected fertilizers. Progressive Fish-Culturist 52:19-25.

Boyd, C. E. 1990. Water Quality in Ponds for Aquaculture. Alabama Agricultural Experiment Station, Auburn University, Alabama.

Buttner, J. K. 1989. Culture of fingerling walleye in earthen ponds–state of the art 1989. Aquaculture Magazine 15(2):37-46.

Buttner, J. K., D. B. MacNeill, D. M. Green, and R. T. Colesante. 1991. Angler associations as partners in walleye management. Fisheries 16(4): 12-17.

Carlander, K. D., and P. M. Payne. 1977. Year-class abundance, population, and production of walleye *(Stizostedion vitreum vitreum)* in Clear Lake, Iowa, 1948-74, with varied fry stocking rates. Journal of the Fisheries Research Board of Canada 34:1792-1799.

Cheshire, W. F., and K. L. Steele. 1972. Hatchery rearing of walleyes using artificial food. Progressive Fish-Culturist 34:96-99.

Clouse, C. P. 1991. Evaluation of Zooplankton Inoculation and Organic Fertilization for Pond Rearing Walleye Fry to Fingerlings. Master's thesis, Iowa State University, Ames, Iowa.

Conover, M. C. 1986. Stocking cool-water species to meet management needs. Pages 31-39 *in* R. H. Stroud, ed. Fish Culture in Fisheries Management. American Fisheries Society, Fish Culture Section and Fisheries Management Section, Bethesda, Maryland.

Dobie, J. 1956. Walleye pond management in Minnesota. Progressive Fish-Culturist 18:51-57.

Dupree, H. K., and J. V. Huner. 1984. Third Report to the Fish Farmers–The Status of Warmwater Fish Farming and Progress in Fish Farming Research. United States Fish and Wildlife Service, Washington, D.C.

Ellison, D. G., and W. G. Franzin. 1992. Overview of the symposium on walleye stocks and stocking. North American Journal of Fisheries Management 12:271-275.

Forney, J. L. 1976. Interactions between yellow perch abundance, walleye predation, and survival of alternate prey in Oneida Lake, New York. Transactions of the American Fisheries Society 103:15-24.

Fox, M. G. 1989. Effect of prey density and prey size on growth and survival of juvenile walleye *(Stizostedion vitreum vitreum)*. Canadian Journal of Fisheries and Aquatic Sciences 46:1323-1328.

Fox, M. G., and D. D. Flowers. 1990. Effect of fish density on growth, survival, and production of juvenile walleyes in rearing ponds. Transactions of the American Fisheries Society 119:112-121.

Fox, M. G., J. A. Keast, and R. J. Swainson. 1989. The effect of fertilization regime on juvenile walleye growth and prey utilization in rearing ponds. Environmental Biology of Fishes 26:129-142.

Geiger, J. G., C. J. Turner, K. Fitzmayer, and W. C. Nichols. 1985. Feeding habits of larval and fingerling striped bass and zooplankton dynamics in fertilized rearing ponds. Progressive Fish-Culturist 47:213-223.

Glamazda, V. V., and Y. A. Katretskiy. 1980. Trophic relationships between zooplankton and bacterioplankton in Tsimlyansk reservoir. Hydrobiological Journal 22:19-21.

Harding, L. M. 1991. Evaluation of Pond Management Practices to Enhance Production of Fingerling Walleye. Master's thesis, Iowa State University, Ames, Iowa.

Jahn, L. A., L. M. O'Flaherty, G. M. Quartucci, J. H. Kim, and X. Mao. 1989. Analysis of Culture Ponds to Enhance Fingerling Production. Completion Report, Federal Aid Project IDCF-50-9, Illinois Department of Conservation, Springfield, Illinois.

Kindschi, G. A., and F. T. Barrows. 1991. Evaluation of an electronic counter for walleye fry. Progressive Fish-Culturist 53:180-183.

Knud-Hansen, C.F., T. R. Batterson, C. D. McNabb, I.S. Harahat, K. Sumantadinata, and H. M. Eidman. 1991. Nitrogen input, primary productivity and fish yield in fertilized freshwater ponds in Indonesia. Aquaculture 94:49-63.

Li, S., and G. G. Ayles. 1981. An Investigation of Feeding Habits of Walleye *(Stizostedion vitreum vitreum)* Fingerlings in Constructed Earthen Ponds in the Canadian Prairies. Canadian Technical Report of Fisheries and Aquatic Sciences 1040, Department of Fisheries and Oceans, Winnipeg, Manitoba.

Li, S., and J. A. Mathias. 1982. Causes of high mortality among cultured larval walleyes. Transactions of the American Fisheries Society 111:710-721.

Ludwig, G. M., and D. L. Tackett. 1991. Effects of using rice bran and cottonseed meal as organic fertilizers on water quality, plankton and growth and yield of striped bass, *Morone saxatilis,* fingerlings in ponds. Journal of Applied Aquaculture 1(1):79-94.

Makarewicz, J. C., and G. E. Likens. 1979. Structure and function of the zooplankton community of Mirror Lake, New Hampshire. Ecological Monographs 49:109-127.

Mathias, J. A., and S. Li. 1982. Feeding habits of walleye larvae and juveniles: comparative laboratory and field studies. Transactions of the American Fisheries Society 111:722-735.

NRC (National Research Council). 1982. United States-Canadian Tables of Feed Composition, 3rd rev. National Academy of Sciences, Washington, D.C.

Oliver, D. R. 1971. Life history of the Chironomidae. Annual Review of Entomology 16:211-230.

Piper, R. G., I. B. McElwain, L. E. Orme, J. P. McCraren, L. G. Fowler, and J. R. Leonard. 1982. Fish Hatchery Management. U.S. Fish and Wildlife Service, Washington, D.C.

Qin, J., and D. A. Culver. 1992. The survival and growth of larval walleye, *Stizostedion vitreum,* and trophic dynamics in fertilized ponds. Aquaculture 108:257-276.

Rees, R. A., and S. F. Cook. 1983. Evaluation of optimum stocking rate of striped bass × white bass fry in hatchery rearing ponds. Proceedings of the Southeastern Association of Fish and Wildlife Agencies 37:257-266.

Richard, P. D., and J. Hynes. 1986. Walleye Culture Manual. Ontario Ministry of Natural Resources, Fish Culture Section, Toronto, Ontario.

SAS Institute, Inc. 1985. SAS User's Guide: Statistics. SAS Institute, Inc., Gary, North Carolina.

Schroeder, G.L. 1978. Autotrophic and heterotrophic production of microorganisms in intensely-manured fish ponds, and related fish yields. Aquaculture 14:303-325.

Smith, L. L., and J. B. Moyle. 1943. Factors influencing production of yellow pike-perch, *Stizostedion vitreum vitreum*, in Minnesota rearing ponds. Transactions of the American Fisheries Society 73:243-261.

Steel, R. G. D., and J. H. Torrie. 1980. Principles and Procedures of Statistics, 2nd ed. McGraw-Hill Book Company, New York, New York.

Summerfelt, R. C., C. P. Clouse, and L. M. Harding. 1993. Pond production of fingerling walleye, *Stizostedion vitreum*, in the northern Great Plains. Journal of Applied Aquaculture 2(3/4):33-58.

Swanson, G. M., and F. J. Ward. 1985. Growth of juvenile walleye, *Stizostedion vitreum vitreum* (Mitchill), in two man-made ponds in Winnipeg, Canada. Internationale Vereinigung für Angewandte Limnologie Verhandlungen 22:2502-2507.

Thurston, R. V., R. C. Russo, and K. Emerson. 1979. Aqueous Ammonia Equilibrium–Tabulation of Percent Un-ionized Ammonia. EPA-600/3-79-091, Environmental Research Laboratory-Duluth, U.S. Environmental Protection Agency, Duluth, Minnesota.

Lethal Effects of Elevated pH and Ammonia on Early Life Stages of Hybrid Striped Bass

David L. Bergerhouse

ABSTRACT. Hybrid striped bass larvae (striped bass, *Morone saxatilis*, ♀ × white bass, *M. chrysops*, ♂) are often stocked into fertile culture ponds. High rates of photosynthesis may result in elevated pH which can be lethal to fry and can affect the toxicity of ammonia. Six-hour static toxicity tests were performed on hybrid larvae of various ages to determine the toxicity of elevated pH and the effects of elevated pH on ammonia toxicity. Six-hour mortality-threshold pH's with no measurable ammonia were estimated for various age larvae and found to be between pH's 9.8 and 10.2 for D2 (D1 is day of hatch), between 9.0 and 9.4 for D4, between 8.8 and 9.2 for D13, and between 9.2 and 9.4 for D20 fish. The addition of 0.7% NaCl had no effect on the toxicity of elevated pH to D2 or D4 fish but caused a significant reduction in the mortality of D13 and D20 fish exposed to high pH. The tolerance to elevated pH decreased as ammonia concentration increased. Sub-lethal un-ionized ammonia concentrations increased the toxicity of elevated pH, suggesting an interaction of pH and un-ionized ammonia toxicity.

David L. Bergerhouse, Cooperative Fisheries Research Lab and Department of Zoology, Southern Illinois University Carbondale, Carbondale, IL 62901-6511, USA.

[Haworth co-indexing entry note]: "Lethal Effects of Elevated pH and Ammonia on Early Life Stages of Hybrid Striped Bass." Bergerhouse, David L. Co-published simultaneously in the *Journal of Applied Aquaculture*, (The Haworth Press, Inc.) Vol. 2, No. 3/4, 1993, pp. 81-100; and: *Strategies and Tactics for Management of Fertilized Hatchery Ponds* (ed: Richard O. Anderson and Douglas Tave) The Haworth Press, Inc., 1993, pp. 81-100. Multiple copies of this article/chapter may be purchased from The Haworth Document Delivery Center [1-800-3-HAWORTH; 9:00 a.m. - 5:00 p.m. (EST)].

81

INTRODUCTION

The culture of hybrids between the striped bass, *Morone saxatilis*, and white bass, *M. chrysops*, is of increasing interest both for production of food fish and for stocking to enhance sport fishing. The dependable production of hybrid striped bass fingerlings is vital to both of these endeavors. Pond production of fingerling hybrids has been highly variable, with some ponds yielding good numbers of fingerlings, while others yield few or none. High rates of photosynthesis can cause periods of elevated pH in ponds. Elevated pH can be lethal to fish and can increase the proportion of ammonia in the toxic un-ionized (NH_3) form. Periods of elevated pH at or near the time fry are stocked into culture ponds may be one factor contributing to inconsistent production. It is, therefore, important to determine the tolerance of hybrid striped bass fry to elevated pH and to determine the effects of elevated pH on the toxicity of ammonia.

High pH has been shown to be toxic to fish by several investigators. Trama (1954) considered the upper lethal pH value for bluegill, *Lepomis macrochirus*, to be 10.5. Jordan and Lloyd (1964) reported 50% mortality in rainbow trout, *Oncorhynchus mykiss*, fingerlings at a pH of 9.86, though roach, *Rutilus rutilus*, fingerlings survived for ten days at pH 10.15. The upper lethal pH value for brook trout, *Salvelinus fontinalis*, was estimated to be 9.8 after acclimation to pH 6.8 (Daye and Garside 1975). Bergerhouse (1992) reported 6-hour mortality-threshold pH values to be between pH's 10.0 and 10.3 for age D4 (D1 is day of hatch) walleye, *Stizostedion vitreum*, and between pH's 9.8 and 10.0 for both D9 and D13 walleye. The tolerance of hybrid striped bass fry to elevated pH has not been adequately investigated.

The uptake of CO_2 by vegetation during periods of high rates of photosynthesis may deplete the supply of free CO_2 and cause a shift in the equilibrium of the carbonate buffer system (Wetzel 1975). Hydroxyl ions can accumulate, resulting in a pH increase if the buffering system is inadequate and the supply of bicarbonate becomes depleted (Cole 1983).

Values of pH high enough to be potentially toxic to fish have been reported by investigators in water bodies of various types. Spring and summer pH values as high as 10.0 were reported by

LaJeone (1972) in Tower Lake, a warm, nutrient-rich cooling lake in Southern Illinois. Values of pH as high as 11.2 were reported by Lewis and Bergerhouse (1988) in an abandoned power plant cooling canal converted for production of sport fish fingerlings. High pH values and a resulting fish kill were reported by Jordan and Lloyd (1964) in the River Tweed in Great Britain. Morris (1988) reported surface pH values as high as 10.1 in hatchery ponds in Mississippi. These reports suggest that peaks in pH due to photosynthesis, may be a fairly common occurrence.

Ammonia exists in solution in equilibrium between the ionized form (NH_4) and the un-ionized form (NH_3). Ionized ammonia is essentially non-toxic to fish (Sheehan and Lewis 1986). Un-ionized ammonia is highly toxic to fish, with acute toxicity levels generally falling between 0.5 and 3 mg NH_3-N/L (Robinette 1976; Thurston et al. 1981; Broderius et al. 1985; Sheehan and Lewis 1986). The proportion of the total ammonia in the toxic un-ionized form increases as the pH and temperature increase (Emerson et al. 1975). Elevated pH can, therefore, be directly lethal to fry and can also increase the toxicity of ammonia.

This study used short-term toxicity tests to examine the acute lethal effects of elevated pH and ammonia on hybrid striped bass. It was designed to determine the value at which pH becomes toxic and the concentration of ammonia that affects that toxicity. The relationship of a larval fish to its environment changes rapidly as it undergoes developmental changes. Therefore, the tolerance of hybrid striped bass of several ages to elevated pH and ammonia was evaluated.

MATERIALS AND METHODS

Striped bass ♀ × white bass ♂ hybrid larvae that had hatched on 26 April, 1988 were received from Arkansas Aquatics, Keo, Arkansas on 27 April. They were held in screened, aerated refuse containers set in indoor raceways in the fish lab at the Quad Cities Nuclear Station, Cordova, Illinois until they reached age D6. They were then released into the raceway, and hourly feeding with brine shrimp nauplii was begun. The raceways were supplied with heated well water with an ambient pH between 8.0 and 8.5. Temperature was maintained between 18 and 19.5°C.

Experiments were performed using hybrid striped bass from four age groups: D2, D4-D5, D13-D14, and D20. Six-hour static toxicity tests at various pH values between 8.0 and 10.75 and at ammonia concentrations between 0 and 13.1 mg TAN (total ammonia nitrogen)/l were performed to determine the toxicity of elevated pH values and ammonia concentrations. A series of toxicity tests with and without ammonia was performed in a salt solution to determine the effects of added salt on pH-induced and ammonia-induced toxicity. A solution of 0.7% NaCl was used to provide an environment that was approximately isotonic with fish tissues.

Toxicity tests were performed in clear plastic containers available commercially as disposable cocktail glasses. The containers were 5.5-cm diameter at the bottom, 8.8-cm diameter at the top, and 7.0-cm maximum depth. Each container was filled with 200 ml of the appropriate test solution and placed in a water bath for the test period. The water bath was flushed continually with 18°-19.5°C water. Three replicate groups of 10 fry each were used for each test condition, though difficulty in counting small fry led to some variation in the number of test animals. To minimize handling stress, the fry were never removed from the water during transfer to the test container. Larvae were drawn into a clear glass tube with a 4-mm inside diameter, counted, and expelled into the test container. The amount of water transferred with the fry was <0.5% of test solution volume to avoid changing the pH or total ammonia concentration.

Test solutions were mixed using heated well water which had no measurable ammonia. The pH value was adjusted by titration with 1.0 and 0.1 N NaOH using an Orion[1] model 501 digital ionalyzer with an Orion 91-04 combination pH electrode. Enough test solution for the three replicates of a given experimental condition was adjusted to the target pH value in a single batch and then distributed to the test containers. Ammonia solutions were prepared by dissolving ammonium sulfate in water prior to pH adjustment. Total ammonia concentration was checked by means of a Corning Ammonia Combination Electrode. Adequate ammonia solution for three replications each of a series of pH values was prepared as a batch of several liters. This solution was divided into jars, the pH value in

---

1. Use of trade names does not imply endorsement.

each jar was adjusted to the appropriate pH with NaOH, and the solution from each jar distributed to three test containers. Standard TAN concentrations tested were 0.26, 1.3, 2.6, 7.8, and 13.1 mg TAN/L.

Mortality in the test containers was checked hourly, and times and numbers of dead larvae were recorded. Data from the three replications of a given treatment were combined to form a single group with n = 30 for analysis. The proportion dead was calculated for each time period, from which the standard error of the proportion was calculated and 95% confidence intervals were determined. Mortalities of two different treatments at a given time period were considered to be significantly different at $\alpha = 0.05$ if the 95% confidence intervals of the two treatments at that time did not overlap. Toxicity tests of any given treatment were performed over a range of pH values. The lowest pH value in the series represented no pH adjustment and was, therefore, considered a pH control group. Mortality curves that were significantly different from this pH control group were considered to have significant pH-related mortality. The pH control groups for pH trials with no added ammonia or salt were considered to be controls for combined effects.

RESULTS AND DISCUSSION

The mortality of D2 hybrid striped bass fry at pH 9.8 was not significantly different from the control group, reaching about 16.7% by the end of the 6-hour test period (Figure 1). Mortality at pH 10.2 was significantly different and greater than mortality of the control group, reaching 67.7% by the end of the test period. A 6-hour mortality-threshold pH value, the lowest pH value at which significant mortality occurred, was therefore estimated to be between pH values of 9.8 and 10.2 for D2 hybrid striped bass fry.

Older hybrid striped bass fry were considerably less tolerant of elevated pH than D2 fry. The 6-hour mortality-threshold pH dropped to between 9.0 and 9.4 at age D4, was between 8.8 and 9.2 at age D13, and between 9.2 and 9.4 at age D20. Striped bass larvae have been noted to have a sensitive period between age D3 and D5, during which the stress of handling or shipping results in high mortality (Lewis et al. 1981). The lower tolerance to elevated pH of

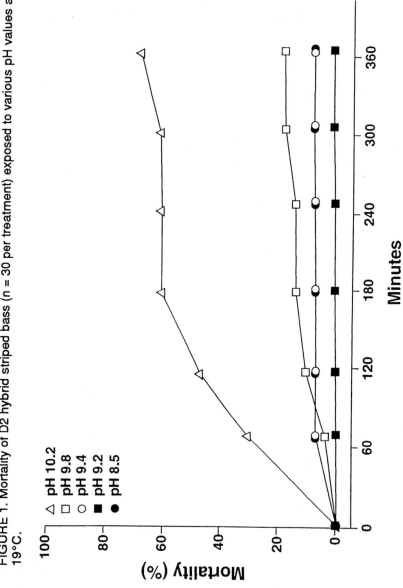

FIGURE 1. Mortality of D2 hybrid striped bass (n = 30 per treatment) exposed to various pH values at 19°C.

hybrid striped bass fry age D4 and older was likely due to some developmental change or group of changes which occurred by age D4. Daye and Garside (1976) reported histopathologic changes in various tissue of fingerling brook trout exposed to high or low pH values. They reported injury to the gill and to the integument of the operculum above pH 9.0, injury to the cornea and lens of the eye above pH 9.5, injury to the naris and the body integument above pH 10.0, and injury to the esophagus above pH 10.8. The development of gill tissue and dependence on gill respiration, coupled with the high sensitivity of gill tissue to elevated pH, may explain the decreased tolerance of older hybrid striped bass to elevated pH values.

Hybrid striped bass were considerably less tolerant of elevated pH than other sportfish species. Six-hour mortality-threshold pH values for hybrid striped bass age D4 and older were considerably lower than similar threshold values reported for walleye; northern pike, *Esox lucius*; and channel catfish, *Ictalurus punctatus* (Figure 2). The low tolerance of hybrid striped bass to elevated pH is of interest, considering the variable success of fingerling production in ponds. High photosynthetic rates in ponds can result in pH peaks well above the pH values that caused significant mortality. Even a relatively brief pH peak may cause significant or complete mortality of young hybrid striped bass. The variable nature of hybrid striped bass production in ponds may be due, in part, to their high sensitivity to elevated pH values.

The sensitivity of hybrid striped bass to elevated pH increased sharply with the addition of even a small amount of ammonia to the test container. The addition of 0.26 mg TAN/L (un-ionized ammonia concentration = 0.04 mg NH_3-N/L) resulted in significant mortality at a pH value of 8.75 for D5 fry (Figure 3); total mortality reached 40% by the end of the 6-hour bioassay period. The 6-hour mortality-threshold pH value for D5 hybrid striped bass with 0.26 mg TAN/L added was estimated to be between 8.38 and 8.75. The 6-hour mortality threshold pH values decreased as ammonia concentration increased over a range of total ammonia concentrations from 0.26 to 13.1 mg TAN/L (Figure 4). D14 fish were somewhat more tolerant of ammonia than were D5 larvae (Figure 5), but ammonia tolerance decreased by age D20 (Figure 6).

There appears to be an interaction between the lethal effects of

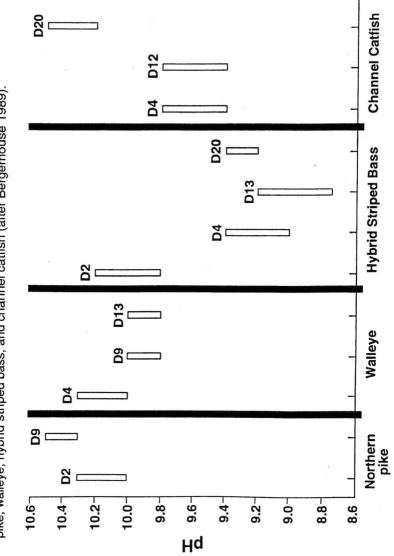

FIGURE 2. Comparison of the 6-hour mortality-threshold pH values of fry of various ages of northern pike, walleye, hybrid striped bass, and channel catfish (after Bergerhouse 1989).

88

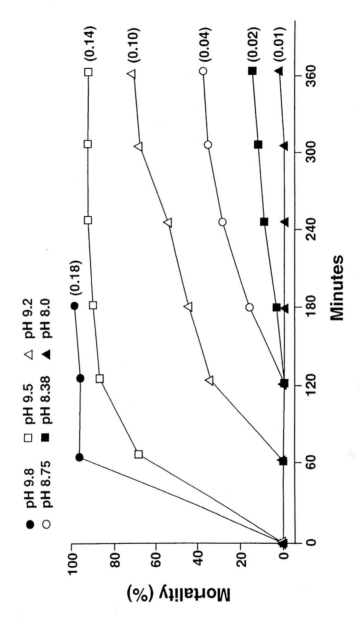

FIGURE 3. Mortality of D5 hybrid striped bass (n = 30 per treatment) exposed to various pH values at 19°C with 0.26 mg TAN/L added. Un-ionized ammonia concentrations (mg NH_3-N/L) are in parentheses.

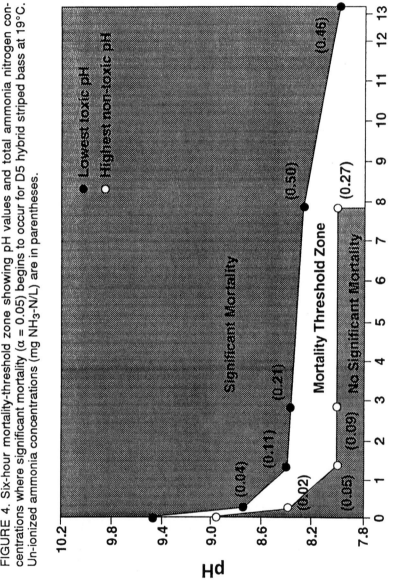

FIGURE 4. Six-hour mortality-threshold zone showing pH values and total ammonia nitrogen concentrations where significant mortality ($\alpha = 0.05$) begins to occur for D5 hybrid striped bass at 19°C. Un-ionized ammonia concentrations (mg NH_3-N/L) are in parentheses.

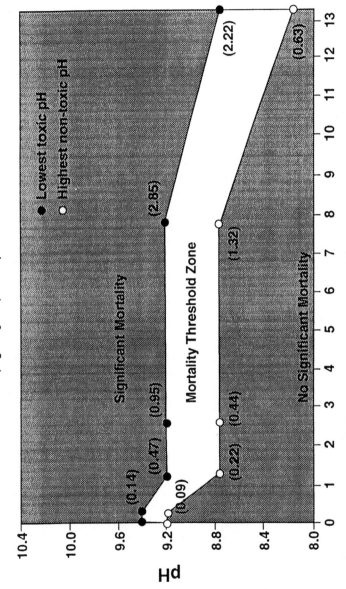

FIGURE 5. Six-hour mortality-threshold zone showing pH values and total ammonia nitrogen concentrations where significant mortality ($\alpha = 0.05$) begins to occur for D14 hybrid striped bass at 19°C. Un-ionized ammonia concentrations (mg NH_3-N/L) are in parentheses.

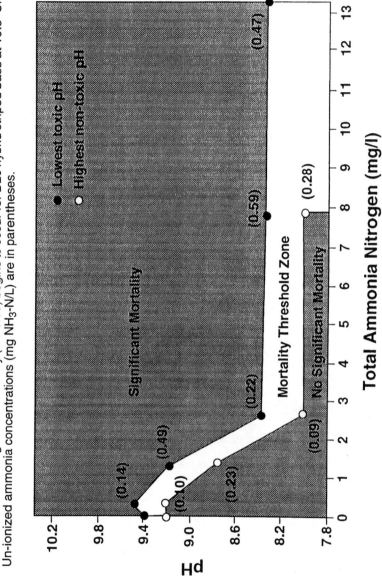

FIGURE 6. Six-hour mortality-threshold zone showing pH values and total ammonia nitrogen concentrations where significant mortality ($\alpha = 0.05$) begins to occur for D20 hybrid striped bass at 19.5°C. Un-ionized ammonia concentrations (mg NH_3-N/L) are in parentheses.

elevated pH and ammonia beyond a simple increase in the proportion of the total ammonia in the toxic un-ionized form. Significant mortality of D14 larvae occurred at an un-ionized ammonia concentration of 0.47 mg NH_3-N/L at a pH value of 9.2 (Figure 5). The un-ionized ammonia concentration alone cannot be responsible for this mortality since a higher un-ionized ammonia concentration (1.32 mg NH_3-N/L) at pH 8.75 resulted in no significant mortality within the 6-hour bioassay period. Similarly, the pH value alone cannot be responsible for this mortality, since a pH of 9.2 with a lower un-ionized ammonia concentration (0.09 mg NH_3-N/L) resulted in no significant mortality. Some interaction must occur between the effects of un-ionized ammonia and elevated pH to cause mortality. A similar interaction effect was reported for walleye fry (Bergerhouse 1992). Thurston et al. (1981) showed an increase in sensitivity to un-ionized ammonia for fathead minnows, *Pimephales promelas*, and rainbow trout above pH 9.0. Tomasso et al. (1980) reported 24-hour un-ionized ammonia LC_{50} values of channel catfish fingerlings that were highest (least toxic) at pH 8.0, lowest (most toxic) at pH 7.0, and intermediate at pH 9.0. Sheehan and Lewis (1986) demonstrated that isotonic dehydration accounted for the apparent increase in toxicity of un-ionized ammonia at lower pH values. They suggested that the combined effects of un-ionized ammonia and osmotic stress explained the apparent increase in un-ionized ammonia toxicity as pH values decreased. Results presented here suggest a potentiation effect between un-ionized ammonia toxicity and the toxicity of high pH. Such an interaction would account for the apparent increase in un-ionized ammonia toxicity above pH 9.0 observed by Tomasso et al. (1980) and Thurston et al. (1981).

The addition of 0.7% NaCl had no effect on the toxicity of elevated pH to D2 and D4 hybrid striped bass with no ammonia added. Mortality was significantly reduced by the addition of salt for D13 larvae (Figure 7), and no mortality of D20 larvae occurred at pH's as high as 10.2 when 0.7% NaCl was added. Salt is often used by fish culturists to reduce mortality caused by stress or injury in situations where fish are being handled. Lewis and Lewis (1971) demonstrated that damage to the integument of freshwater fish can cause a reduction in blood osmolality, leading to mortality. The pH-induced damage to tissues reported by Daye and Garside (1976)

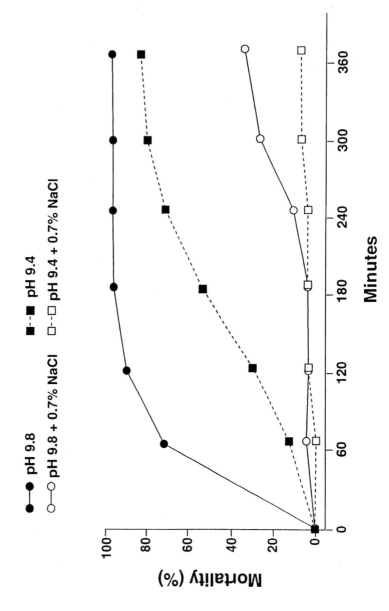

FIGURE 7. Effects of 0.7% NaCl on the mortality of D13 hybrid striped bass (n = 30 per treatment) exposed to elevated pH values at 19°C.

94

may allow an influx of water and subsequent loss of ions from tissues. A saline medium may reduce the loss of ions from the tissues and, therefore, affect pH-related mortalities. D2 hybrid striped bass fry were more tolerant to elevated pH than were older fry, presumably due to the lack of well-developed gills which are more sensitive to pH-related damage than other tissues. The lack of gills may preclude any beneficial effect of added NaCl, since the gills are likely the primary site of water influx. Hybrid striped bass age D4 and older were less tolerant of elevated pH than were D2 larvae, presumably due to some developmental changes including gill development. The addition of NaCl did not increase pH tolerance in D4 larvae but caused drastic increases in pH tolerance for D13 and D14 fry. Well-developed gills in older fish may make them more subject to damage due to high pH but allow some osmotic protection from the effects of high pH in a salt solution.

The addition of salt generally reduced mortality in toxicity tests with ammonia. The mortality of D5 hybrid striped bass exposed to 1.3 mg TAN/L was significantly reduced by the addition of 0.7% NaCl at pH values of 9.5 and below but not at pH 9.8 (Figure 8). Mortalities of D14 (Figure 9) and D20 hybrids exposed to 1.3 mg TAN/L were significantly reduced by the addition of salt at pH 9.8 and below, which includes all pH values tested.

The proportion of the total ammonia in the un-ionized form decreases as the ionic strength increases (Solderberg and Meade 1991). The reduction in the concentration of un-ionized ammonia in toxicity tests with added NaCl may contribute to the apparent mitigating effect of salinity on ammonia toxicity. The values for un-ionized ammonia shown in Figures 8 and 9 with added salt were calculated as described by Solderberg and Meade (1991) to account for the effects of increased ionic strength. The reductions in the concentrations of un-ionized ammonia do not appear to be adequate to account for the magnitude of the decreases in mortality. These results suggest a mechanism of ammonia-related mortality involving osmotic balance. Lloyd and Orr (1969) demonstrated an increase in the absorption of water by rainbow trout exposed to ammonia. They associated a threshold LC_{50} for ammonia toxicity with a urine flow rate that was six times the normal flow rate. An ammonia-induced diuresis and subsequent loss of sodium from the tissues could con-

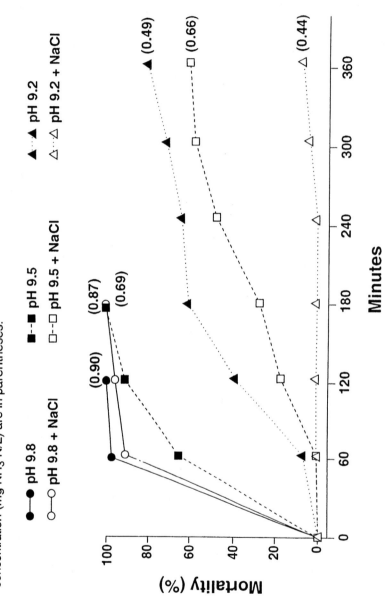

FIGURE 8. Effects of 0.7% NaCl on the mortality of D5 hybrid striped bass (n = 30 per treatment) exposed to ammonia (1.3 mg TAN/L) at various elevated pH values at 19°C. Un-ionized ammonia concentration (mg NH_3-N/L) are in parentheses.

● pH 9.8   ■—■ pH 9.5   ▲···▲ pH 9.2
○—○ pH 9.8 + NaCl   □—□ pH 9.5 + NaCl   △···△ pH 9.2 + NaCl

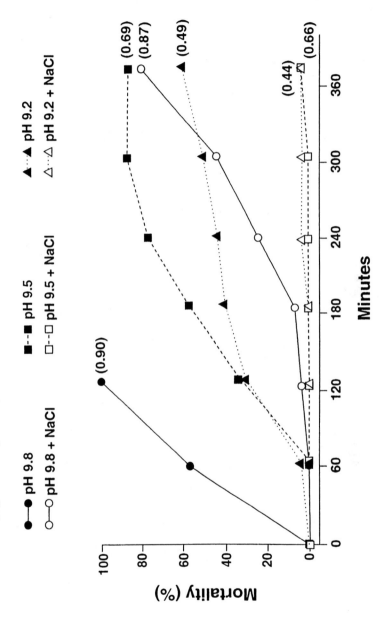

FIGURE 9. Effects of 0.7% NaCl on the mortality of D14 hybrid striped bass (n = 30 per treatment) exposed to ammonia (1.3 mg TAN/L) at various elevated pH values at 19°C. Un-ionized ammonia concentration (mg NH3-N/L) are in parentheses.

97

tribute to ammonia-induced mortality of hybrid striped bass fry. An aqueous environment with 0.7% NaCl would reduce the osmotic gradient, thereby reducing or delaying mortality. Herbert and Shurben (1965) reported that the toxicity of ammonia to rainbow trout decreased with increasing salinity up to about 30% seawater, which is approximately isotonic with the blood. The apparent interaction of the lethal effects of ammonia and elevated pH may be due to the combined stress from the two factors. However, the fact that the toxic effects of both elevated pH and ammonia are moderated in a saline environment suggests that both factors affect osmotic balance. Loss of ions from tissues due to ammonia-induced diuresis, combined with loss of ions due to water influx through pH damaged tissues, would result in more rapid mortality than from either factor alone.

The lethal effects of elevated pH should be recognized as a potential factor contributing to the variable success of hybrid striped bass production in ponds. The potentiation effects of elevated pH and un-ionized ammonia preclude the establishment of independent threshold limits for the two factors. The combined effect must be considered in determining whether water quality in a given water body is suitable for the survival of hybrid striped bass fry.

ACKNOWLEDGMENTS

I thank William M. Lewis for his many contributions towards the completion of this work. In addition I thank Commonwealth Edison Co., Chicago, Illinois for the use of facilities in conjunction with the Quad Cities Spray Canal Project.

REFERENCES

Bergerhouse, D.L. 1989. Lethal Effects of Elevated pH and Ammonia on Early Life Stages of Several Sportfish Species. Doctoral dissertation, Southern Illinois University Carbondale, Carbondale, Illinois.

Bergerhouse, D.L. 1992. Lethal effects of elevated pH and ammonia on early life stages of walleye. North American Journal of Fisheries Management 12:356-366.

Broderius, S., R. Drummond, J. Fiandt, and C. Russom. 1985. Toxicity of ammonia to early life stages of smallmouth bass at four pH values. Environmental Toxicology and Chemistry 4:87-96.

Cole, G.A. 1983. Textbook of Limnology. Mosby, St. Louis, Missouri.

Daye, P. G., and E. T. Garside. 1975. Lethal levels of pH for brook trout *Salvelinus fontinalis* (Mitchill). Canadian Journal of Zoology 53:639-641.

Daye, P. G., and E. T. Garside. 1976. Histopathologic changes in surficial tissues of brook trout *Salvelinus fontinalis* (Mitchill), exposed to acute and chronic levels of pH. Canadian Journal of Zoology 54:2140-2155.

Emerson, K., R. C. Russo, R. E. Lund, and R. V. Thurston. 1975. Aqueous ammonia equilibrium calculations: Effect of pH and temperature. Journal of the Fisheries Research Board of Canada 32:2379-2383.

Herbert, D. W. M., and D. G. Shurben. 1965. The susceptibility of salmonid fish to poisons under estuarine conditions. II. Ammonium chloride. International Journal of Air and Water Pollution 9:89-91.

Jordan, D. H. M., and R. Lloyd. 1964. The resistance of rainbow trout (*Salmo gairdneri* Richardson) and roach (*Rutilus rutilus* L.) to alkaline solutions. International Journal of Air and Water Pollution 8:405-409.

LaJeone, L. J. 1972. The Physical Limnology of Tower Lake, Madison County, Illinois. Master's thesis, Southern Illinois University Edwardsville, Edwardsville, Illinois.

Lewis, W. M., and D. L. Bergerhouse. 1988. Investigations of the Potential for Producing Sport Fishes of Various Sizes at the Quad Cities Spray Canal. Progress report January 1-December 31, 1987. Fisheries Research Laboratory, Southern Illinois University Carbondale, Carbondale, Illinois.

Lewis, S. D., and W. M. Lewis. 1971. The effect of zinc and copper on the osmolality of blood serum of the channel catfish, *Ictalurus punctatus* Rafinesque, and golden shiner, *Notemigonus crysoleucas* Mitchill. Transactions of the American Fisheries Society 100:639-643.

Lewis, W. M., R. C. Heidinger, and B. L. Tetzlaff. 1981. Tank Culture of Striped Bass. Illinois Striped Bass Project, IDC F-26-R. Fisheries Research Laboratory, Southern Illinois University Carbondale, Carbondale, Illinois.

Lloyd, R., and L. D. Orr. 1969. The diuretic response by rainbow trout to sublethal concentrations of ammonia. Water Research 3:335-344.

Morris, J. E. 1988. Role of Artificial Feeds and Feeding Regimes on the Culture of Hybrid Striped Bass Fry. Doctoral dissertation, Mississippi State University, Mississippi State, Mississippi.

Robinette, H. R. 1976. Effect of selected sublethal levels of ammonia on the growth of channel catfish (*Ictalurus punctatus*). Progressive Fish-Culturist 38:26-29.

Sheehan, R. J., and W. M. Lewis. 1986. Influence of pH and ammonia salts on ammonia toxicity and water balance in young channel catfish. Transactions of the American Fisheries Society 115:891-899.

Solderberg, R. W., and J. W. Meade. 1991. The effects of ionic strength on un-ionized ammonia concentration. Progressive Fish-Culturist 53:118-120.

Thurston, R. V., C. R. Russo, and G. A. Vinogradov. 1981. The effect of pH on the toxicity of the un-ionized ammonia species. Environmental Science and Technology 15:837-840.

Tomasso, J. R., C. A. Goudie, B. A. Simco, and K. B. Davis. 1980. Effects of environmental pH and calcium on ammonia toxicity in channel catfish. Transactions of the American Fisheries Society 109:229-234.

Trama, F. B. 1954. The pH tolerance of the common bluegill (*Lepomis macrochirus* Rafinesque). Notulae Naturae (Phila.) 256(Feb):1-13.

Wetzel, R. G. 1975. Limnology. Saunders, Philadelphia, Pennsylvania.

Apparent Problems and Potential Solutions for Production of Fingerling Striped Bass, *Morone saxatilis*

Richard O. Anderson

ABSTRACT. Data on production of fingerling striped bass, *Morone saxatilis*, in 103 ponds at 17 hatcheries in 1988 were related to information on management practices and water qualities. Production was highly variable, with a median of 50,000/ha. No survival was evident in 12% of ponds; numbers harvested exceeded 250,000/ ha in 5% of ponds. Best survival resulted when larvae were stocked within 2 days after pond filling was started and at hatcheries with relatively low application rates of fertilizers during the prestocking interval. No survival was evident in ponds with pH >9.0 in the week after stocking. High pH's and probably toxic concentrations of un-ionized ammonia were related to water being in ponds for too long prior to stocking and to high applications of fertilizers. Examination of samples of larvae collected at stocking and the week after stocking documented a problem with swim bladder inflation. When larvae were stocked at D5 or D6 (D1 is day of hatch), inflation success averaged 91%; when stocked at ≥D7, inflation success averaged 60%. Measures proposed to improve production include: (1) Stock larvae prior to the critical period of swim bladder inflation

Richard O. Anderson, National Fish Hatchery and Technology Center, United States Fish and Wildlife Service, 500 East McCarty Lane, San Marcos, TX 78666, USA. Correspondence may be addressed to 3618 Elms Court, Missouri City, TX 77459, USA.

[Haworth co-indexing entry note]: "Apparent Problems and Potential Solutions for Production of Fingerling Striped Bass, *Morone saxatilis.*" Anderson, Richard O. Co-published simultaneously in the *Journal of Applied Aquaculture*, (The Haworth Press, Inc.) Vol. 2, No. 3/4, 1993, pp. 101-118; and: *Strategies and Tactics for Management of Fertilized Hatchery Ponds* (ed: Richard O. Anderson, and Douglas Tave) The Haworth Press, Inc., 1993, pp. 101-118. Multiple copies of this article/chapter may be purchased from The Haworth Document Delivery Center [1-800-3-HAWORTH; 9:00 a.m. - 5:00 p.m. (EST)].

101

(D4-D5). (2) Start filling ponds as close to stocking as is feasible (filling can be completed after stocking). (3) If the concentration of available nitrogen is low, apply no more than 300 μg/L of nitrogen in the prestocking interval. (4) Add phosphorus as needed, but avoid pH >8.5 until larvae are D14.

INTRODUCTION

An important axiom is that it is hard to find solutions to poorly defined problems. Variable survival and production of phase-I striped bass, *Morone saxatilis,* in fertilized ponds at federal hatcheries led to a workshop on pond management practices in February, 1988. An outcome of that workshop was that hatchery managers agreed to participate in a concerted effort to collect samples of fish and to provide information on management practices, water quality, and fish production to the National Fish Hatchery and Technology Center (NFHTC) in San Marcos, Texas. The objective of this paper is to relate practices and problems based on the information and data reported by the hatchery managers.

MATERIALS AND METHODS

Study in 1988

Managers at 16 national fish hatcheries (NFH) and one state fish hatchery (SFH) provided information on pond filling, fertilizing, and stocking practices. The location and names of the study hatcheries were: Alabama–Carbon Hill NFH; Florida–Welaka NFH; Georgia–Warm Springs NFH, Bo Ginn NFH; Louisiana–Natchitoches NFH; Mississippi–Meridian NFH, Private John Allen NFH; North Carolina–Edenton NFH, McKinney Lake NFH; Ohio–Senecaville SFH; Oklahoma–Tishomingo NFH; South Carolina–Orangeburg NFH; Texas–Inks Dam NFH, San Marcos NFHTC, Uvalde NFH; Virginia–Harrison Lake NFH. Production data for striped bass were provided from 103 ponds. Each hatchery was provided with detailed instructions on water quality measurements that were to be obtained. Water temperature, dissolved oxygen, and pH were measured in the morning and sometimes in the afternoon. Water quality data for some ponds at some hatcheries were unreported. A 0.5-m diameter icthyoplankton net was provided to each hatchery to collect samples of larval fish. Fish samples were collected at stocking and on one or more nights in the week after stocking. A light was used to attract larvae, which were caught by a vertical lift of the icthyoplankton net. Preserved samples were shipped to the NFHTC for analysis. Swim bladder inflation of larvae was determined by gross examination with a dissection microscope. Estimates of inflation were based on samples of at least 25 larvae. Fish harvested from ponds were weighed; sample counts were made to estimate numbers. Survival was estimated from numbers reported stocked and harvested.

Hatchery managers made their own decisions on pond filling time, stocking age and density, fertilization materials and application rates, herbicide treatments, harvest schedule, and other management practices. Management programs and schedules were typically site-specific.

Studies in 1989

Larval fish were collected in 1989 at the Natchitoches NFH and McKinney Lake NFH during the week after stocking; a light and dip net were used at night. At the Natchitoches NFH, larvae were received at D1, the day eggs hatched, and held at temperatures >20°C until stocked. Six ponds were stocked on the same day with either D4, D5, or D6 larvae. Four of the ponds were stocked with fry hatched over 2 days. At the McKinney Lake NFH, each of three ponds was stocked with either D6 or D11 larvae when they were received. Swim bladder inflation in larval samples from each pond was determined by microscopic examination of preserved fish.

Uvalde NFH used different pond filling schedules and fertilization practices in 1988 and 1989. Management practices and fish production for the two years were compared.

Correlation analyses relating survival and harvest density to management practices were performed with Harvard Graphics 3.0 software. Probability values were based on tables in Zar (1984).

RESULTS AND DISCUSSION

Swim Bladder Inflation

The age of fish stocked in 1988 ranged from D5 to D15. In ponds stocked with D5 or D6 larvae, inflation success was more consistent and averaged 91% (Figure 1); in ponds stocked with D7-D15 larvae, inflation success was highly variable and averaged only 60%. These data suggest that the critical inflation period occurred at <D7 and that conditions are more favorable for inflation in ponds than in hatchery tanks.

In 1989 at the Natchitoches NFH, there was an apparent inverse relationship between the age stocked and percentage of inflation: D4-D5–95%; D5–85%; D5-D6-28-75%; D6–33%. Lower inflation was evident for larvae held for a longer period. When larvae were held at >20°C, the critical inflation period probably occurred around D5 or D6.

In 1989 at the McKinney Lake NFH, the averages (and ranges) of inflation for larvae stocked at the two ages were: D6–77% (40-100%); D11–55% (45-68%). The wide range for D6 is puzzling. The critical period of inflation for an individual fish may last less than 24 hours. The critical age for this stage of development is probably inversely related to water temperature after hatching. The

FIGURE 1. Swimbladder inflation (%) in samples of striped bass larvae collected from individual ponds at study hatcheries in 1988. Inflation was 100% in four ponds stocked at D5.

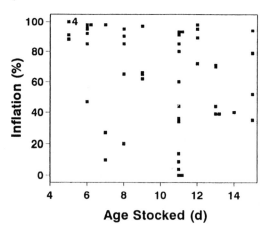

best age to stock to increase inflation success may be D4 or D5. A 99% inflation success was observed when D4 striped bass larvae were stocked at the NFHTC in 1991 (Anderson 1993b).

Striped bass without inflated swim bladders are more vulnerable to stress than normal fish. Low inflation percentage might be expected to contribute to low fingerling survival. However, no relationship between inflation and survival to harvest was apparent in 1988 (Figure 2).

Management Practices in 1988

Total alkalinity of the water sources at the 17 hatcheries ranged from 4 to 380 mg/L CaCO$_3$ (Table 1). Both surface waters and wells were used to fill ponds. Both sources provided some hatcheries with considerable concentrations of nitrate and ammonia nitrogen (N) and soluble reactive or total phosphorus (P). The concentration of N for 11 hatcheries for which some data were available ranged from 70 to 9,000 μg/L. Only three hatcheries had data for soluble reactive P; concentrations ranged from 6 to 120 μg/L. The concentrations of total P, including organic sources, at four other hatcheries ranged from <10 to 140 μg/L.

Hatchery managers used a wide variety of organic and chemical fertilizers (Table 2). Twelve hatcheries used some combination of both

FIGURE 2. Survival (%) of striped bass as a function of swimbladder inflation at study hatcheries in 1988.

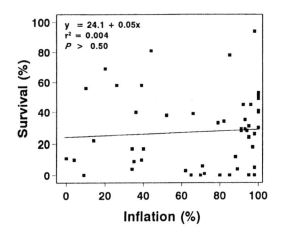

TABLE 1. Water sources for study hatcheries producing striped bass; S or W refer to surface or well. Total alkalinity and hardness are expressed as mg/L $CaCO_3$; phosphorus reported as either soluble reactive or as total phosphorus.

Hatchery	Source	Alk.	Hard.	NO_3^--N (mg/l)	NH_3-N (mg/l)	P (μg/l)
1	W	20	20	8.80	0.20	-
2	S	174	184	0.50	1.23	-
3	S	111	174	0.07	0.00	6
4	S	4	22	-	-	-
5	S	380	380	0.64	0.16	-
6	W	134	24	-	-	-
7	S	52	52	0.12	0.12	140*
8	S	-	5	-	-	-
9	W	286	94	2.00	0.95	-
10	S	138	210	-	-	-
11	W	250	300	1.40	0.02	12
12	S	12	21	0.19	0.05	90*
13	W	100	94	-	0.20	-
14	S	150	195	0.10	0.05	<10*
15	S	80	91	0.46	0.04	70*
16	W	216	518	1.17	0.08	120
17	S	10	30	-	-	-

*Total = Soluble Reactive P + Organic P

types (Table 3); four hatcheries used only organic fertilizers, and one hatchery used only chemical fertilizers from the time of initial filling through the first week after stocking. Several hatcheries added prepared fish food daily during the latter part of the production period.

Six of the hatcheries treated problem growths of aquatic plants with

TABLE 2. Fertilizer materials and chemicals applied to striped bass ponds at study hatcheries. The code numbers are used to report the organic fertilizers (1-11), chemical fertilizers (1-10), herbicides (1-2), and other treatments (1-3) at individual hatcheries in Table 3. Number of hatcheries using each item is in parentheses.

	Fertilizers		Treatments	
Code	Organic	Chemical	Herbicides	Other
1	Cottonseed meal (8)	33-0-0 (0)	Copper sulfate (4)	Masoten (2)
2	Alfalfa meal, pellets (5)	46-0-0 (4)	Aquazine (4)	Oil (5)
3	Soybean meal (3)	0-20-0 (2)		$KMnO_4$ (1)
4	Peanut meal (1)	0-46-0 (5)		
5	Meat scrap meal (3)	0-54-0 (0)		
6	Wheat shorts (1)	10-34-0 (3)		
7	Fish food (2)	11-37-0 (2)		
8	Brewers yeast (1)	18-17-0 (1)		
9	Peanut hay (1)	10-10-10 (1)		
10	Hay, other (5)	18-46-0 (2)		
11	Chicken manure (1)			

TABLE 3. Production of striped bass and management practices at study hatcheries in 1988. Order was based on average survival; bust ponds had survival <10%. Prestock is the period ponds had water prior to stocking. Harvest density and yield are averages; number/ha is in thousands. Ages S and H are averages at stocking and at harvest. Prestock fertilization rates were based on total application divided by weeks when applications were made for more than 7 days; poststock rates were for the week after stocking. Fertilizer and treatment codes are listed in Table 2.

Hatchery	Survival (%)	Ponds No.	Ponds Bust	Harvest (no./ha)	Harvest (kg/ha)	Prestock (d)	Age S	Age H	Fertilizers (Codes) Organic	Fertilizers (Codes) Chem	Fertilization rates (µg/l/week) Prestock N	Prestock P	Poststock N	Poststock P	Treatments (Codes) Herb.	Other
1	66	5	0	161.8	113.0	2	7	43	1	–	0	0	1,113	496	–	–
2	53	5	0	208.2	48.9	5	6	34	–	4,7	311	513	0	197	–	–
3	46	5	0	115.6	78.9	12	15	58	1,8,10	8	1,212	452	245	35	1	–
4	45	5	0	145.3	63.9	7	9	48	5,10,11	3	3,452	1,911	0	0	1,2	–
5	44	6	0	163.7	62.7	0	5	38	2,5,7	6	1,320	88	876	799	1,2	1,2
6	44	4	0	154.9	28.6	16	6	30	1,10	–	151	27	626	59	–	2
7	36	5	0	112.8	51.3	4	5	46	4,7,9	–	966	133	489	36	–	2
8	29	5	0	73.5	27.6	18	8	33	3	6	1,887	320	364	17	2	2,3
9	21	5	3	46.1	16.3	6	11	48	2	2	1,593	799	662	44	2	–
10	19	4	0	47.5	16.9	6	7	37	2,3	3	800	183	400	150	–	–
11	16	14	6	16.4	12.2	14	11	38	1,6	–	1,216	56	759	36	–	–
12	6.7	5	4	18.3	20.6	29	5	44	3	6,10	2,625	759	2,786	672	–	–
13	5.5	3	2	14.9	11.5	43	12	53	1,2,10	4,6	824	1,090	625	380	1	2
14	5.1	2	1	7.9	7.0	7	10	43	10	4	*	*	70	841	–	–
15	1.2	4	4	2.0	1.2	7	6	37	1,5	2	2,200	918	720	433	–	–
16	0	3	3	0	0	12	6	20	1,2,6	2,4,7	1,169	654	74	5	–	1
17	0	4	4	0	0	32	9	18	1	2,4,9	6,713	1,282	961	44	–	–

*Rye grass crop flooded.

either $CuSO_4$ or Aquazine™[1] (Table 3). Treatments for problem invertebrates included Masoten, fuel oil, or $KMnO_4$ at six hatcheries. Some hatcheries had populations of fairy shrimp, *Streptocephalus* sp., or clam shrimp, *Cyzicus* sp., which were untreated. One hatchery drained and refilled ponds prior to stocking, in order to reduce densities of fairy shrimp.

There was a wide range of time intervals between when filling was started and when ponds were stocked. At one hatchery, filling began the day larvae were stocked and continued for 14 days after stocking. This schedule was used so that larvae could feed on the initial hatch of fairy shrimp and clam shrimp. At another hatchery, water was in ponds an average of 43 days prior to stocking. At eight hatcheries, the interval from starting to fill ponds to stocking fish was ≤7 days. The average age of fish stocked ranged from D5 to D15; the age at seven hatcheries was D5 or D6. The average age of fish at harvest ranged from D30 to D58.

Production and Survival

The estimated density of larvae stocked in ponds ranged from 100,000 to 500,000/ha. Most ponds were stocked at a density of 200,000 to 300,000/ha. The number harvested was highly variable. Median harvest was about 50,000/ha; 12% of ponds had no survival, while 5% of ponds had a harvest ≥ 250,000/ha (Figure 3). There was no relationship between age stocked and number harvested or survival (Figure 4).

Survival was >25% in all 13 ponds at the three hatcheries where larvae were stocked within 2 days after pond filling was started (Figure 5); survival was ≤ 10% in 40% of the ponds that were stocked ≥6 days after pond filling was started. A long period between when pond filling was started and stocking, in order to allow higher densities of zooplankton, failed to enhance survival. Based on these observations, pond filling should be started as close to stocking as is feasible.

Not all hatcheries provided data on pH. Almost all of the measurements reported were made in the morning. There was no survival in all seven ponds when the maximum pH measured to D14

---

1. Use of trade names does not imply endorsement.

FIGURE 3. Frequency distribution (%) of number of striped bass harvested from individual ponds at study hatcheries in 1988.

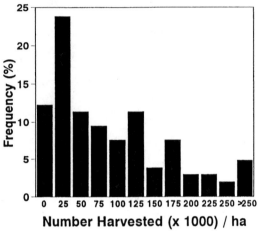

was above 9.0 (Figure 6). High buffering capacity and alkalinity >200 mg/L as $CaCO_3$ did not prevent pH >9.0 when ponds were filled 2 weeks prior to stocking. High pH and low survival may have been evident in more ponds if pH had been measured in the afternoon. Low survival of young larvae at high pH was consistent with the results reported by Bergerhouse (1993).

Average weekly application rates of N and P for the periods prior to stocking and the first week after stocking were estimated, based on reported pond volumes, quantities of fertilizers used (Table 3), and estimates of N and available P (Anderson 1993a). When the fertilization period prior to stocking was >7 days, total quantities added were divided by the time interval in weeks. The N and P content of organic meals was assumed to be available during the week that the fertilizers were applied. When hay was added, it was assumed that N and P were available at a rate of 25%/week. There was a pattern of higher average survival in ponds with lower application rates of N and P (Figure 7). High application rates of N probably resulted in higher concentrations of total ammonia nitrogen (TAN); high application rates of P probably stimulated excessive photosynthetic rates and caused elevated pH.

A short interval between pond filling and stocking combined

FIGURE 4. Density of striped bass harvested (A) and percent survival (B) from individual ponds at study hatcheries in 1988 as a function of age stocked.

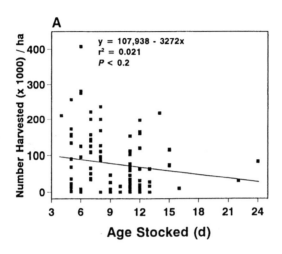

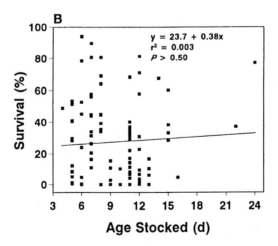

FIGURE 5. Survival (%) of striped bass in individual ponds at study hatcheries in 1988 as a function of the interval in days from when pond filling was started to when larvae were stocked.

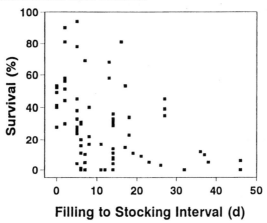

FIGURE 6. Survival (%) of striped bass in individual ponds at study hatcheries in 1988 as a function of maximum pH recorded to D14.

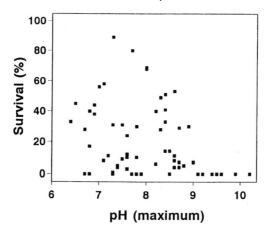

with low to moderate application rates of N and P, especially when larvae are <D14, should be the rule for the production of striped bass. The concentration of N available each week might best be <600 μg/L as used by Culver et al. (1993) when the N that is applied is in the form of protein, urea, or ammonium compounds. Only three

FIGURE 7. Average survival (%) of striped bass at study hatcheries in 1988 as a function of average weekly application rate (mg/L) of nitrogen (A) and phosphorus (B) during the interval prior to stocking. The open symbol was omitted from statistical analyses.

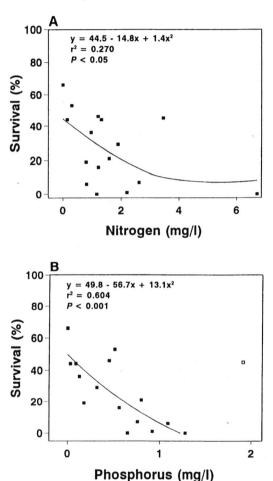

hatcheries used a weekly application rate for N of ≤ 311 μg/L during the interval prior to stocking; these hatcheries ranked 1, 2, and 6 in survival. One hatchery with no survival used an average weekly rate of 6,713 μg/L. A weekly application of no more than 300 μg/L TAN may be best until larvae are D14; phosphorus should be added at a weekly rate that is low enough to avoid pH >8.5 during this time interval. The best amount of N and P to add may vary from time to time, hatchery to hatchery, and pond to pond.

Uvalde NFH 1989

Uvalde NFH was one of the hatcheries where there was no survival of striped bass in 1988. However, they had excellent production in 1989. Harvest density and yield in three of four study ponds in 1989 exceeded 250,000/ha and 100 kg/ha. In 1988, pond filling was started 14 days prior to stocking. The total amount of N added as organic and chemical fertilizers was 570 and 2,810 μg/L, respectively. The total amount of P added as organic and chemical fertilizers was 20 and 2,020 μg/L, respectively. Maximum pH in ponds the week after stocking was 9.1-10.2; Secchi disk transparency was 30-40 cm.

In 1989, the hatchery received eggs, hatched them, and held larvae under controlled water temperatures <19°C until they were stocked 2 days after pond filling was started. The average amounts of organic N added prior to stocking and the first week after stocking were 1,520 and 720 μg/L, respectively; inorganic N amounts added during these periods were 150 and 160 μg/L, respectively. Average amounts of organic P added prior to stocking and the first week after stocking were 150 and 160 μg/L, respectively; inorganic P amounts added during these intervals were 190 and 310 μg/L, respectively. Maximum pH during the week after stocking ranged from 8.4 to 8.7; survival ranged from 52 to 73%. It is logical to conclude that the short interval between filling and stocking, as well as the much lower amounts of inorganic N and P added in 1989, avoided the problems of excessive photosynthesis and high pH and improved survival.

The well water used to fill ponds at Uvalde NFH had concentrations of NO_3^-–N of 1,170 μg/L and soluble reactive P of 120 μg/L. This water had N and P concentrations that were much higher than those proposed by Culver et al. (1993); at this hatchery, it should be unnecessary to add any nutrients to ponds prior to stocking and to be able to delay fertilization for a few days after stocking.

Successful production of fingerling striped bass has been an elusive goal at many hatcheries. Problems influencing success have not been well defined. Striped bass are much more sensitive to stress from adverse environmental factors and normal hatchery practices than most other species that are propagated. As a result, hatchery managers have less margin for error in decisions on "what to do" and "when to do it."

When to start filling ponds is an important management decision. According to Geiger and Turner (1990), pond filling should be initiated 13 days prior to stocking. However, Brewer and Rees (1990) reported that ponds for striped bass were filled 5-10 days prior to stocking at many hatcheries. Braschler (1975) recommended that pond filling should be started after fry are received at D1 or D2 in order to delay development of predatory insects. Based on the 1988 data, an interval of 5 days or more between starting to fill and stocking ponds may reduce survival. At most hatcheries, it may be best to start filling ponds no more than 2 days prior to stocking and to complete filling after larvae are stocked.

What age larvae to stock is an important decision. Until procedures are worked out to ensure swim bladder inflation in hatchery tanks, stocking D4 or D5 larvae is a logical practice. Hatcheries that produce fingerling striped bass should have the facilities to hatch eggs or hold larvae prior to stocking. The receiving hatchery should have the capability of holding larvae in water of good quality at a temperature of 15-19°C. Fish should be well adapted to the water quality and temperature of ponds when stocked.

The best density of larvae to stock should be related to pond productivity and predicted survival. Better management of pond filling and fertilization may result in more ponds with potential harvests of $\geq 250,000$/ha and yields of 100 kg/ha.

Several federal hatcheries had water supplies with poor buffering capacities and alkalinities <30 mg/L $CaCO_3$. Some hatcheries applied agricultural lime to ponds. The best way to manage productivity of waters with low alkalinity has not been determined. Dissolved inorganic carbon can be the primary limiting factor for algal production in fertilized ponds with low alkalinity (McNabb et al. 1990). Organic fertilizers may be well suited to such systems since they provide a source of carbon, as well as N and P. The risk of high pH should be less with organic than with chemical nutrients (Anderson 1993b).

The results reported by Barkoh and Rabeni (1990) provide good evidence to support the use of alfalfa meal rather than cottonseed meal as an organic fertilizer. Results reported by Ludwig and Tackett (1991) and Mims et al. (1993) show that rice bran is also an effective organic fertilizer. Densities of 10,000 *Daphnia magna*/L have been produced in laboratory cultures, and cultures have been sustained for >5 months solely on rice bran (DePauw et al. 1981). Rice bran is usually the cheapest of all organic meals used as pond fertilizers. Physical properties of alfalfa meal and rice bran are better than those of cottonseed meal. If a slurry of cottonseed meal is mixed in water, most quickly settles as a heavy sludge. Of the three materials, rice bran has the smallest particle sizes; consequently, more is likely to stay in suspension longer and be consumed directly by zooplankton. It also has a fat content of 13%, compared to about 2% for alfalfa meal and 0.5-3.7% for cottonseed meal (Hubbell 1989). The higher energy content of rice bran may be beneficial to zooplankton.

Whatever organic material is used, the amount applied and the frequency of application should take into account the resulting available N and P, as well as the potential biochemical oxygen demand. A conservative approach of more frequent applications of smaller amounts can be an advantage in case of cloudy weather.

At hatcheries with well-buffered water supplies and an alkalinity ≥ 100 mg/L $CaCO_3$, there is a considerable cost advantage in using chemical sources of N and P (Anderson 1993b). Urea is probably a safer source of N than ammonium nitrate. Even when ammonium nitrate is diluted and evenly applied to ponds, mixing by natural dynamics of pond water will take time. Hot spots of un-ionized ammonia may develop, which could be toxic to sensitive larvae (Piotrowska-Opuszynska 1984). Urea is probably converted to ammonia after an interval in which complete mixing of ponds is likely (C. E. Boyd, Auburn University, pers. comm.). Phosphoric acid is an effective and low cost source of P (Murad and Boyd 1987).

Research is needed to evaluate the kinetics of N and P in hatchery ponds. Information on the rate of conversion of urea to ammonia would help managers to decide on how much and how often it should be applied. The fraction of total N and P in organic meals that is available for biotic production should also be evaluated.

There is a need and an opportunity to improve the success, as well as ecological and economic efficiencies, in the production of striped bass and many other fingerling fishes. Well-managed, fertilized ponds are likely to be the production method of choice for most species needed for aquaculture or for fishery management.

ACKNOWLEDGMENTS

Data from the study hatcheries in 1988 and 1989 would have been unavailable without the cooperation and support of the hatchery managers and their staffs. Many people at the NFHTC contributed to sample analyses and data recording. The initial reporting of data for individual hatcheries was completed by Stewart Leon. Staff at the NFHTC were also essential for preparation and editorial review of this paper. C. Berkhouse provided invaluable assistance and support in the preparation of figures and statistical analyses; the patience and skills of M. Estrada passed all the tests in the production of tables and revisions of the text.

REFERENCES

Anderson, R. O. 1993a. New approaches for management of fertilized hatchery ponds. Journal of Applied Aquaculture 2(4):000-000.

Anderson, R. O. 1993b. Effects of organic and chemical fertilizers and biological control of problem organisms on production of fingerling striped bass, *Morone saxatilis*. Journal of Applied Aquaculture 2(4):000-000.

Barkoh, A., and C. F. Rabeni. 1990. Biodegradability and nutritional value to zooplankton of selected organic fertilizers. Progressive Fish-Culturist 52:19-25.

Bergerhouse, D. L. 1993. Lethal effects of elevated pH and ammonia on early life stages of hybrid striped bass. Journal of Applied Aquaculture 2(3/4):000-000.

Braschler, E. W. 1975. Development of pond culture techniques for striped bass *Morone saxatilis* (Walbaum). Proceedings of the Southeastern Association of Game and Fish Commissioners 28:44-48.

Brewer, D. L., and R. A. Rees. 1990. Pond culture of phase I striped bass fingerlings. Pages 99-120 *in* R. M. Harrell, J. H. Kerby, and R. V. Minton, eds. Culture and Propagation of Striped Bass and its Hybrids. Striped Bass Committee, Southern Division, American Fisheries Society, Bethesda, Maryland.

Culver, D. A., S. Madon, and J. Qin. 1993. Percid pond production techniques: Timing, enrichment, and stocking density manipulation. Journal of Applied Aquaculture 2(3/4):000-000.

DePauw, N., P. Laureys, and J. Morales. 1981. Mass cultivation of *Daphnia magna* Straus on rice bran. Aquaculture 25:141-152.

Geiger, J. G., and C. J. Turner. 1990. Pond fertilization and zooplankton management techniques for production of fingerling striped bass and hybrid striped bass. Pages 79-98 *in* R. M. Harrell, J. H. Kerby, and R. V. Minton, eds. Culture and Propagation of Striped Bass and its Hybrids. Striped Bass Committee, Southern Division, American Fisheries Society, Bethesda, Maryland.

Hubbell, C. H. 1989. Feedstuffs analysis tables. Feedstuffs, February 20:36.

Ludwig, G. M., and D. L. Tacket. 1991. Effects of using rice bran and cottonseed meal as organic fertilizers on water quality, plankton, and growth and yield of striped bass, *Morone saxatilis,* fingerlings in ponds. Journal of Applied Aquaculture 1(1):79-94.

McNabb, C. D., T. R. Batterson, B. S. Premo, C. F. Knud-Hansen, H. M. Eidman, C. K. Lin, K. Jai Yen, J. E. Hansen, and R. Chuenpagdee. 1990. Managing fertilizers for fish yield in tropical ponds in Asia. Pages 169-172 *in* R. Hirano and I. Hanyu, eds. The Second Asian Fisheries Forum. Asian Fisheries Society, Manila, Philippines.

Mims, S. D., J. A. Clark, J. C. Williams, and D. B. Rowe. 1993. Comparisons of two by-products and a prepared diet as organic fertilizers on growth and survival of larval paddlefish, *Polyodon spathula,* in earthen ponds. Journal of Applied Aquaculture 2(3/4):000-000.

Murad, A., and C. E. Boyd. 1987. Experiments on fertilization of sport-fish ponds. Progressive Fish-Culturist 49:100-107.

Piotrowska-Opuszynska, W. 1984. The influence of nitric fertilizers on physico-chemical conditions in nursery ponds. Roczniki Nauk Rolniczych H-100, 4:111-132. (in Polish; English summary)

Zar, J. H. 1984. Biostatistical Analysis. Prentice-Hall, Inc. Englewood Cliffs, New Jersey.

Effects of Organic
and Chemical Fertilizers
and Biological Control
of Problem Organisms on Production
# of Fingerling Striped Bass,
Morone saxatilis

Richard O. Anderson

ABSTRACT. Sixteen 0.04-ha ponds were fertilized with similar amounts of nitrogen (N) and available phosphorus (P) provided by either alfalfa meal, urea and phosphoric acid, or a combination where half of the P was provided by phosphoric acid and half by alfalfa meal. Half of the ponds fertilized with the combination of nutrients were stocked with adult male common carp, *Cyprinus carpio*, at an average biomass of 168 kg/ha to provide biological control of rooted aquatic plants and clam shrimp, *Cyzicus morsie*. All ponds were stocked with 25,000 larval striped bass, *Morone saxatilis*, at an age of D4 (D1 is the day of hatch). Median harvest density and survival were about 100,000/ha and 16%. Number harvested was directly related to numbers sampled with a light and dip net at D5 and D8. Low survival was probably related to high afternoon water temperatures (25-26°C) and relatively low morn-

Richard O. Anderson, National Fish Hatchery and Technology Center, United States Fish and Wildlife Service, 500 East McCarty Lane, San Marcos, TX 78666, USA. Correspondence may be addressed to 3618 Elms Court, Missouri City, TX 77459, USA.

[Haworth co-indexing entry note]: "Effects of Organic and Chemical Fertilizers and Biological Control of Problem Organisms on Production of Fingerling Striped Bass, *Morone saxatilis*." Anderson, Richard O. Co-published simultaneously in the *Journal of Applied Aquaculture*, (The Haworth Press, Inc.) Vol. 2, No. 3/4, 1993, pp. 119-149; and: *Strategies and Tactics for Management of Fertilized Hatchery Ponds* (ed: Richard O. Anderson and Douglas Tave) The Haworth Press, Inc., 1993, pp. 119-149. Multiple copies of this article/chapter may be purchased from The Haworth Document Delivery Center [1-800-3-HAWORTH; 9:00 a.m. - 5:00 p.m. (EST)].

119

ing dissolved oxygen (4.6-6.5 mg/L) when larvae were D5. Stocking larvae at an age prior to swim bladder inflation resulted in an inflation success of 99%. Dynamics of average net photosynthesis, chlorophyll *a* concentrations, and densities of crustacean zooplankton, as well as mean number, biomass, and length of fingerlings harvested were similar in all treatments, regardless of whether the source of P was organic, inorganic, or the combination. Growth rate of larvae from D5 to D8 was considered satisfactory (≥ 0.4 mm/day) with average densities of crustacean zooplankton of 10-20/L. The average growth rate of larvae from D8 to D40 was negatively related to number harvested. A weekly fertilization rate of available P from 28-38 μg/L resulted in satisfactory average growth rate (0.83 mm/day) of decreasing numbers of larvae at increasing ages: D8-D19–150,000/ha, D19-D25–125,000/ha, and D25-D40–73,000/ha. The presence of adult common carp provided several benefits: effective control of *Chara* and filamentous algae; a lower average percentage of fingerlings stranded in vegetation when ponds were drained (0.4% vs 10.1%); lower pH; effective control of clam shrimp.

INTRODUCTION

Questions that produce a wide range of answers from hatchery managers are: What kinds of fertilizer do you use in your ponds? What frequency and rate of fertilization do you use? A variety of materials and rates was used at hatcheries producing fingerling striped bass, *Morone saxatilis,* studied in 1988; too much fertilizer prior to stocking ponds was related to low survival (Anderson 1993b). The primary purpose of this study was to compare the responses of plants, zooplankton, and striped bass to moderate amounts of nitrogen (N) and phosphorus (P) applied as either alfalfa meal, as urea and phosphoric acid, or in combination. A secondary objective of the study was to evaluate the use of common carp, *Cyprinus carpio,* as a method of biological control of rooted aquatic plants and clam shrimp, *Cyzicus morsie.*

For the production of percid fingerlings at hatcheries in Ohio, Culver et al. (1993) recommended a weekly fertilization rate to achieve concentrations of available N and P of 600 and 30 μg/L, respectively. These recommendations deviated from published guidelines for propagation of striped bass. Geiger and Turner (1990) recommended an aggressive fertilization program with both organic and inorganic fertilizers during the 13-day period prior to stocking. Concentrations proposed can be estimated, assuming a pond mean depth of 0.9 m and values of N (6.6%) and P (0.3%) in cottonseed meal (Anderson 1993a). The Geiger and Turner (1990) average weekly rates prior to stocking ponds ranged from 1,566 to 4,530 μg N/L and from 326 to 1,225 μg P/L. These rates are 2.6-40 times higher than the rates proposed by Culver et al. (1993).

The recommendations made by Geiger and Turner (1990) are conservative when compared to those made by Piper et al. (1982), who proposed an initial application of 2,470 kg chicken manure, 1,729 kg alfalfa meal, 494 kg meat meal, and 124 kg super phosphate/ha. If average pond depth is assumed to be 0.9 m and N and P values are similar to those presented by Anderson (1993a), these materials add up to 5,919 μg N/L and 2,451 μg P/L. Fish were to be stocked 3-5 days after applying this load. A second application might be added in 30-40 days.

MATERIALS AND METHODS

On March 5, 1991, sixteen 0.04-ha earthen ponds at the National Fish Hatchery and Technology Center (NFHTC), San Marcos, Texas were filled to about 75% of full volume in order to cover most of the bottom. During the next 2 to 4 days, they were treated with one or two applications of anhydrous ammonia. Total ammonia nitrogen (TAN) concentrations the day after treatment ranged from 28 to 70 mg/L. The objectives of this treatment were to kill viable filamentous algae, the initial hatch of fairy shrimp, *Streptocephalus* sp., clam shrimp, and crayfish, *Procambarus* sp. All ponds were drained on April 2.

On April 11, the inside concrete drain kettles were filled with water. Filling was resumed on April 14 and was interrupted when ponds were 50% full on April 17. Crustacean zooplankton were harvested from two non-study ponds, as described by Graves and Morrow (1988b), and stocked into study ponds on April 17, 18, and 19 at a total density of 1.6/L.

On the evening of April 17, four lots of striped bass larvae that had been hatched 0030-0330 on April 15 at the A. E. Wood State Fish Hatchery, San Marcos, Texas were transferred to the NFHTC.

Larvae had been held at 17-19°C. Larvae from each lot were equally divided into 16 plastic garbage cans; water temperature was 19°C. Fish were tempered in the cans prior to stocking by periodically adding pond water over a 7-hour period. Water temperatures were 24-25°C when fish were released at about 0400 on April 18. Larvae looked normal and were active when stocked. An estimated 25,000 D4 larvae (D1 is the day of hatch) were stocked in each pond.

Filling was resumed during the day on April 18, and ponds were 90% full on April 23 (D9). Filling continued gradually after April 23 through metered valves at the shallow ends of ponds. Constant water levels were maintained with float valves.

Ponds were initially fertilized on April 19 (D5). Four ponds were fertilized with alfalfa meal at a rate that increased with age of fish. This strategy was based on the logic that pond productivity should increase with increasing food needs of fish. A relatively low initial fertilization rate was used to avoid excessive photosynthesis, pH >8.5, and un-ionized ammonia concentrations >50 μg/L until larvae were D14. These values were based on the data of Bergerhouse (1989).

Alfalfa meal was applied in equal amounts to four ponds at weekly rates equivalent to 112, 168, and 224 kg/ha at <D14, D14-20, and D21-D33, respectively. The equivalent concentrations of N in these periods were 285, 427, and 570 μg/L, respectively; equivalent concentrations of available P were 19, 28, and 38 μg/L, respectively. In order to increase the available P during the initial time interval, 7 μg/L of P as phosphoric acid were added to each of the ponds. No other phosphoric acid was used in these ponds after fish were D14. The weekly amount of alfalfa meal was applied in equal increments on Monday, Wednesday, and Friday to avoid a high biochemical oxygen demand. As ponds were filling, the quantity added each day was adjusted to achieve the desired concentration. The alfalfa meal was mixed with water, and the slurry was broadcast into the pond.

Four ponds were fertilized with urea and phosphoric acid. Desired amounts of N were 300 μg/L when larvae were <D14 and 600 μg/L from D14-D33. Desired amounts of available P were 30 μg/L. The amounts of urea and phosphoric acid to add to a pond each week were based on the differences between the desired amount of N or P and the measured concentrations of available N or P in each

pond. Water samples were collected the day prior to fertilization and were analyzed for NO_3–N, TAN, and soluble reactive P. The initial application on April 19 was equivalent to a full week's application for ponds that were 60% full. Subsequent application amounts were based on nutrient analyses made on Sunday. An equal portion of the amount needed for the week was applied on Monday, Wednesday, and Friday. On the last day of fertilization (May 17), the amount added was equal to the amount needed for a full week.

Eight ponds were fertilized with a combination of alfalfa meal, urea, and phosphoric acid to achieve amounts of available N and P similar to those in the ponds that were fertilized only with urea and phosphoric acid. Half of the P was added as alfalfa meal and half as phosphoric acid. When the amount of N needed was not provided by the alfalfa meal, the difference was made up with urea.

Four of the combination study ponds were stocked on April 23 with adult male common carp. The average total weight was 6.7 kg (168 kg/ha). Two of these ponds were selected because they had a history of clam shrimp. The primary objective of stocking common carp was to control the growth of *Chara*. No herbicide or other plant control methods were applied to any of the other ponds. Common carp were also stocked in the expectation that their bottom-feeding activities would increase turbidity, increase the availability of N and P, and provide some control of clam shrimp and other problem invertebrates. Common carp had been treated prior to stocking to control external parasites.

Measurements of dissolved oxygen and water temperature were made daily, usually at 0730 and 1430, at a depth of 0.5 m. The pH was measured every afternoon at the surface with a Whatman[1] pH probe. Secchi disk transparency was also measured in the afternoon. Integrated water samples from the surface to near the bottom were collected with a tube sampler described by Graves and Morrow (1988a) on Sunday and Thursday for analyses of NO_3–N, TAN, soluble reactive P, and chlorophyll *a*. The analyses of NO_3–N and P were made with an Alpkem autoanalyzer model RFA 300. TAN was determined by direct Nesslerization with a Bausch and

---

1. Use of trade names does not imply endorsement.

Lomb Spectronic 20 colorimeter. Chlorophyll *a* was measured as described by Burnison (1980). Turbidity was measured with a Hach Turbidimeter Model 2100A.

Zooplankton were sampled after dark on Monday, Wednesday, and Friday at the deep end of the ponds by means of multiple dips with the tube sampler to collect about 4 L of water. Each sample was concentrated with an 80-μm plankton net and preserved with buffered-sugar-formalin solution (Haney and Hall 1973). Total samples were concentrated; copepods, large cladocerans (*Daphnia*), other cladocerans, and rotifiers were enumerated in a plankton counting wheel.

D5 and D8 striped bass and invertebrates were collected after dark with the aid of a light and a 15- $\times$ 19-cm, fine-mesh dip net and were preserved with formalin. The sampling time at each pond on each night was 5 minutes or less. D19 and D25 striped bass were sampled with a 4-mm mesh fry seine. Samples were also collected when fish were harvested at D37-D40. Total lengths of all fish or a sub-sample of 25 fish (50 at harvest) at each sampling date were measured to the nearest 0.025 mm with a caliper. Magnification was used to measure D5 and D8 fish. Individual fish in harvest samples were blotted with a damp sponge and weighed to the nearest milligram.

Fecal samples were collected from two common carp from each of the two ponds with clam shrimp on April 30 and May 7. Several grams of fecal material were collected from each fish by applying water pressure through a tube inserted into the foregut. The fecal material was examined with a dissection microscope. The fish were restocked, and all common carp were alive at harvest.

Samples of *Chara* and filamentous algae were collected from ponds after they were drained. Four samples were collected from each pond within the 1.2- to 1.5-m contours. Plants were collected from within a square frame with inside dimensions of 0.1 m. The four samples from each pond were pooled, allowed to air-dry for 3 weeks, and then weighed to the nearest gram.

All ponds were drained May 21-24 (D37-D40). Fish stranded during draining were enumerated and included in harvest values. Numbers at harvest were based on estimates of average weight in preserved samples and total weight harvested.

Regression analyses were performed with Harvard Graphics 3.0 software to evaluate relationships among harvest density, total length, growth rate, and zooplankton densities. Levels of significance and ANOVA tests were based on Zar (1984).

RESULTS AND DISCUSSION

Swim Bladder Inflation

No swim bladder inflation was evident in a sample of 200 D4 larvae collected when ponds were stocked. Pond temperatures up to 26°C in the two afternoons following stocking stimulated the rate of development and inflation. Of the 881 D5 larvae collected from 16 ponds, only 10 (1.1%) had no inflation. The success was similar for D8 larvae; only 3 of 373 had no inflation. These results indicate that the probability of inflation success is excellent in ponds when striped bass larvae are stocked prior to the inflation period.

No food was evident in the gut of D5 larvae. Inflation prior to initiation of feeding probably has adaptive significance. Larvae with a swim bladder should be more adept at capturing prey. Reduced growth has been observed for larvae without a swim bladder (unpublished data). These observations conflict with those made by Doroshev (1970) who stated that for most striped bass larvae, inflation occurred at D10-D15 and 1-3 days after the initiation of feeding. Doroshev and Cornacchia (1979) reported that the critical inflation period for striped bass held at 18°C was D6-D8, i.e., the 5th to 7th day after hatching. Bulak and Heidinger (1980) observed inflation as early as D4 for striped bass larvae maintained at 16-18°C.

Nutrient Additions

Well water at the NFHTC contained 1,400 μg/L NO_3–N and 12 μg/L P. Only one pond had a concentration of N <300 μg/L during the period when larvae were <D14 because of the NO_3–N that was added as ponds were being filled. The average NO_3–N concentration in all ponds prior to fertilization on April 18 was 778 μg/L. Average NO_3–N concentrations declined to 238 μg/L on April 21.

When ponds were more than 90% full on April 25, average NO_3^--N concentrations were 101 $\mu g/L$ and ranged from 2 to 289 $\mu g/L$.

TAN concentrations increased from April 18 to April 25; the average concentrations for all ponds were 197 and 270 $\mu g/L$, respectively. On April 25, the average concentration for the treatments ranged from 222 $\mu g/L$ in ponds scheduled to receive urea and phosphoric acid to 348 $\mu g/L$ in ponds stocked with common carp and fertilized with the combination of nutrients.

The average concentrations of available N (NO_3^--N + TAN) declined in all treatments from the time when striped bass were <D14 to the time when they were D21-D33 (Table 1). The average percentages of N as TAN in all ponds in the three intervals were 34, 71, and 83%, respectively. Even though nutrient analyses were conducted only 1 or 2 days after nutrient additions, average concentrations of P were consistently ≤ 5 $\mu g/L$, regardless of fertilizer treatment. Either the uptake of P by plants was rapid or there was a loss by precipitation in the hard water. Based on the concentrations of P and the available N:P ratio determined by the analyses, P was the nutrient limiting photosynthesis in all treatments during the study.

Primary Production

The initial application of nutrients on April 19 stimulated photosynthesis and increased average pH the most in ponds fertilized with phosphoric acid and with the combination of nutrients (Figure 1). Maximum pH was 8.3 on April 21 in two ponds fertilized with phosphoric acid. Estimated concentrations of un-ionized NH_3 on April 21 ranged from 5 to 23 $\mu g/L$. Based on the results presented

TABLE 1. Average concentrations ($\mu g/L$) of NO_3^--N + ammonia-N and soluble reactive phosphorus (P) in each fertilizer treatment during three age intervals of striped bass production (<D14, D14-D20, and D21-D33).

Treatment	<D14 N	<D14 P	D14-D20 N	D14-D20 P	D21-D33 N	D21-D33 P
Organic	598	3	434	4	249	4
Chemical	596	4	343	5	179	6
Combination	530	4	367	3	183	5
Combination + common carp	656	4	554	4	300	5

FIGURE 1. Average pH (A), net photosynthesis (B), and chlorophyll *a* (C) for each treatment. Open triangles indicate fertilization dates.

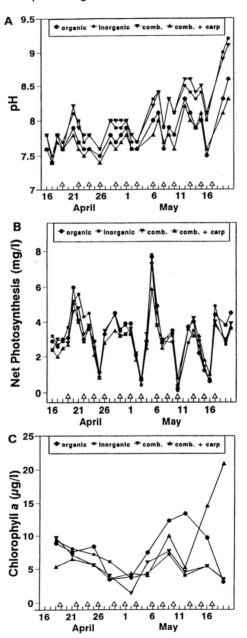

by Bergerhouse (1993), this range of concentrations should be safe for larval striped bass.

There was no obvious trend in pH in any treatment through May 3 (Figure 1); after that, pH trends were generally positive. Declines in pH during the study were usually associated with cloudy weather. The pH observed was consistently higher in ponds fertilized with urea and phosphoric acid and with the combination of nutrients than it was in ponds fertilized with only alfalfa meal or the combination of nutrients with common carp.

On May 1, *Chara* was observed in all ponds except those with common carp. When ponds were drained, the average plant biomass in ponds that had common carp was 66 g/m^2 compared to 340 g/m^2 for the other ponds (Table 2); the difference was significant ($P < 0.05$).

The gain in oxygen concentration from the morning to afternoon is an estimate of net photosynthesis for about 7 hours. From April 16 to 19 prior to fertilization, the average net photosynthesis for each study pond ranged from 2.0 to 3.2 mg/L (Figure 1). There was a consistent and marked increase in photosynthesis during the 2 days after the initial fertilization. Average net photosynthesis for all treatments on April 20 and 21 ranged from 4.0-5.6 mg/L.

The pattern of average net photosynthesis during the study was consistent for all treatments. Low values were associated with cloudy days. The highest values on May 5 were on a sunny afternoon when dissolved oxygen was measured later than usual. There was no consistent difference observed in net photosynthesis among treatments.

TABLE 2. Plant dry weight (g/m^2) for individual 0.04-ha ponds with each treatment. Means of each column followed by the same letter were not significantly different ($P < 0.05$).

	Organic	Chemical	Combination	Combination + common carp
	296	246	336	136
	246	525	427	82
	492	361	437	48
	153	172	412	0
Mean	292a	326a	403a	66b

Oxygen values <5.0 mg/L during the study were observed in only three ponds on 3 days. One of the low readings was in a pond prior to the initial fertilization. Low oxygen was probably related to decomposition of residual terrestrial vegetation. Morning oxygen concentrations were often above saturation. Afternoon values were occasionally >15 mg/L. Under such conditions, estimates of net photosynthesis were conservative, since no correction was made for loss of oxygen to the atmosphere. The values were also conservative since measurements were made in the afternoon when concentrations were probably less than maximum for the day; any gain from dawn to the morning measurement was also unrecorded. Although estimates of net photosynthesis were conservative, measurements at 0730 and 1430 can be used to evaluate responses to fertilization.

Average concentrations of chlorophyll *a* prior to fertilization for the different treatments ranged from 5 to 10 μg/L (Figure 1); after 2 weeks of fertilization, the average concentrations of chlorophyll *a* in all treatments were <5 μg/L. A pond with fairy shrimp had the lowest mean concentration (3.0 μg/L) of chlorophyll *a* of all ponds. Higher average concentrations were evident in the latter part of the study in ponds fertilized with alfalfa meal and in ponds stocked with common carp. Higher concentrations in alfalfa-treated ponds may be related to the weekly addition of P at 38 μg/L, compared to an average addition of about 26 μg/L for the other treatments. Higher average concentrations of chlorophyll *a* observed in ponds with common carp in the last two samples may have been related to the lower abundance of *Chara*.

Zooplankton

The decline in chlorophyll *a* before May 1 may have been related to increasing densities of crustacean zooplankton (Figure 2). Patterns of average zooplankton density observed during the study were similar for all treatments. Maximum average densities of about 100-400/L were observed from April 29 to May 6 (D15-D22) and then declined. When the density of zooplankton on each sampling date from April 26 to May 17 was ranked from low (1) to high (16), the average ranks among treatments were similar, but the ranges within a treatment were wide; the means and ranges were:

FIGURE 2. Average density of crustacean zooplankton for each treatment.

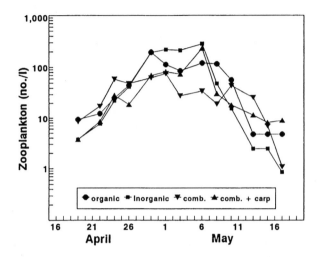

organic–6.9 and 2.0-14.1; chemical–9.8 and 7.0-12.2; combination–7.6 and 5.1-12.4; combination + common carp–8.3 and
4.1-13.0 (Table 3). The pond with the lowest mean rank (2.0) had a
high density of fairy shrimp; the ponds with the next lowest mean
ranks (4.1 and 4.6) were ponds with moderate to high densities of
clam shrimp. The abundance of these shrimp appeared to suppress
the growth of zooplankton populations. Only low numbers of fairy
shrimp or clam shrimp were observed in three of the other ponds.
The average ranks of means in these ponds were 9, 11, and 12.

   Concentrations of chlorophyll a in samples during this study may
not be the best index of feeding conditions or of effects of filter
feeding Cladocera and copepods. The bloom in some ponds included
Volvox, which is too large to be consumed by zooplankton. Samples
from some ponds also included filamentous algae. A better index of
feeding conditions for zooplankton might have been obtained if the
samples had been strained through a plankton net and chlorophyll a
measured from nanoplankton.

   It would be expected that fish predation would influence the
relative rank of zooplankton density. However, there was no apparent relationship observed between fish harvest biomass and mean
rank of zooplankton density. On May 3 (D19) when maximum

TABLE 3. Ranks of density (no./L) of crustacean zooplankton on each sampling date for each pond, mean rank for each pond, overall treatment mean ranks, and ranks of means for each pond. Ranks ranged from low (1) to high (16).

Treatment						Date						Mean of ranks	Rank of means
	4/26	4/29	5/1	5/3	5/6	5/8	5/10	5/13	5/15	5/17		
Organic	11	14	14	13	14	16	16	14	14	15	14.1	16
	16	8	4	5	5	7	7	1	1.5	7.5	6.2	6
	1	1	1	1	1	1	1	2.5	6.5	3.5	2.0	1*
	10	16	11.5	11	9	11	12	12	11.5	7.5	5.4	5
										Mean	6.9	7
Chemical	4	4	2	7	11	14	13	8	5	2	7.0	9
	12	13	16	14	10	8	3.5	10.5	10	3.5	10.0	11
	5	9	11.5	12	12	10	5.5	10.5	11.5	11	9.8	10
	15	15	15	16	13	12	8	9	8		12.2	13
										Mean	9.8	11
Combination	7	5	6.5	4	4	5	10.5	7	3.5	11	6.4	7
	14	6	9	6	8	3.5	2	2.5	9	5	6.5	8
	8	7	6.5	9	6	3.5	3.5	5	1.5	1	5.1	4
	13	11	10	10	7	13	15	16	16	13	12.4	14
										Mean	7.6	11
Combination + common carp	2	3	5	2.5	3	6	5.5	5	6.5	7.5	4.6	3**
	3	2	3	2.5	2	2	10.5	5	3.5	7.5	4.1	2**
	6	12	13	15	15	9	14	15	15	16	13.0	15
	9	10	8	8	16	15	9	13	13	14	11.5	12
										Mean	8.3	8

* pond with highest density of fairy shrimp
** ponds with highest densities of clam shrimp

131

average zooplankton densities were observed, stomach contents of fish collected from 10 of 16 ponds consisted >50% by volume of chironomid larvae; estimated average percentage volume of chironomids in fish stomachs ranged from 11 to 91%. The highest percentage was in the pond with a high density of fairy shrimp and the lowest density of zooplankton. Zooplankton probably would have made a greater contribution to the diet if densities had been higher or if *Daphnia* had been more abundant. Density of large Cladocera during the study exceeded 1/L in only 16 of 208 samples.

Problem invertebrates, as well as fish, were collected with the light and dip net on April 19, eight days after the drain kettles were filled. The number of ponds where different taxa were collected were: Notonectidae–11, Dytiscidae larvae–11, *Chaoborus* larvae–4, Zygoptera nymphs–3. Burleigh et al. (1993) documented how quickly Notonectidae can invade newly flooded ponds. No relationship was apparent between the presence or absence of these invertebrates in samples and survival of striped bass.

Fish Production

The numbers (and density) of fish harvested from each pond (Table 4) ranged from 841 (21,000/ha) to 6,615 (165,400/ha); survival ranged from 3.4 to 26%. Median production and survival were 100,000/ha and 16%. There were no significant differences among treatments in average number, biomass, or size of fish harvested.

There was a significant positive correlation ($r = 0.64$; $P < 0.01$) between the total number of larvae collected with the light and dip net at D5 and D8 and the number harvested (Figure 3). Evaluation of initial survival should be an asset to managers. Ponds with low survival could be drawn down and restocked. Evaluation of initial survival with a light and small dipnet appears to be better than using a tow net (Turner 1988).

Low survival was probably related to afternoon water temperatures, which ranged from 25 to 26°C at D4 and D5. Temperatures above 24°C are considered detrimental (Brewer and Rees 1990). Albrect (1964) observed 1-11% survival of larvae held to D6 at 23.9°C; he found no survival at 26.7°C. Doroshev (1970) reported survival to D6 of 52% at 22-24°C and no survival at 26-27°C.

Lowest harvest numbers (841-2,148) were observed in five

TABLE 4. Harvest numbers and biomass (g) and average total length (mm) and weight (g) of striped bass harvested from each 0.04-ha ponds, and number and percentage of fish stranded during harvest. Means in each column followed by the same letter were not significantly different (*P* < 0.05).

Treatment	Harvest No.	Harvest Wt.	Sizes Ln.	Sizes Wt.	Stranded No.	Stranded %
Organic	1,623	1,688	44	1.04	257	15.8
	5,150	2,117	33	0.41	496	9.6
	3,947	1,062	29	0.27	134	3.4
	4,965	2,135	34	0.43	32	0.6
Mean	3,921a	1,750a	35a	0.54a	230	7.4a
Chemical	5,391	1,170	26	0.22	51	0.9
	4,259	1,576	30	0.37	156	3.7
	1,026	856	42	0.83	331	32.3
	3,690	1,369	32	0.37	17	0.4
Mean	3,592a	1,243a	32a	0.45a	139	9.3a
Combination	4,566	1,443	30	0.32	868	19.0
	5,927	1,233	28	0.21	398	6.7
	4,993	1,353	27	0.27	239	4.8
	841	658	40	0.78	208	24.7
Mean	4,082a	1,172a	31a	0.40a	428	13.8a
Combination	6,615	1,806	29	0.27	14	0.2
+ common carp	5,101	1,413	28	0.27	2	0.0
	1,254	1,376	46	1.10	12	1.0
	2,148	1,143	36	0.53	4	0.2
Mean	3,780a	1,434a	35a	0.54a	8	0.4b

ponds with relatively low oxygen concentrations on the morning of D5; concentrations ranged from 4.6 to 6.5 mg/L compared to 6.6-7.6 mg/L in all other ponds. Since larval striped bass do not have functional gills for respiration until about D10 (R. Heidinger, Southern Illinois University, Carbondale, Illinois, pers. comm.), the combination of high temperature and relatively low oxygen may have been fatal to many larvae.

Since most of the mortality probably occurred shortly after stocking, the numbers harvested are a conservative estimate of the numbers present over the production period. There was a significant negative correlation (r = −0.89; *P* < 0.001) between the average length at harvest and the number harvested (Figure 4). Average lengths in individual ponds for all treatments are both above and below the line. Growth rate and final average length in a pond were more related to density than to treatment.

A total length of 35 mm (≈0.45 g) at D40 for striped bass is a reasonable management objective. Based on the regression in Fig-

FIGURE 3. Relationship between number of striped bass harvested and the total number collected with a light and dip net at D5 and D8.

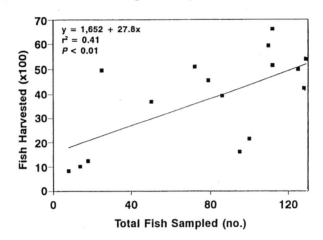

ure 4, this study provided for satisfactory growth of about 84,000 fish/ha. This was well below the 250,000/ha produced in some ponds at study hatcheries in 1988 and 1989 (Anderson 1993b).

The amount of available P each week was apparently too low for good production of phytoplankton, zooplankton, and fish. Based on mean chlorophyll *a* concentrations, the level of productivity in all treatments was too low for most of the production period. Geiger and Turner (1990) suggested that chlorophyll *a* levels should be maintained between 10 and 40 μg/L.

The range of densities of fish and zooplankton provide an opportunity to evaluate interactions. Since fish were sampled at intervals, growth rates can be related to average density of zooplankton and density of fish. Given a target size at D40 of 35 mm and an average length at D5 of 6 mm, average growth rate needs to be 0.83 mm/day.

In the period from D5 to D8 when feeding was initiated, the average density of crustacean zooplankton ranged from 12 to 69/L (Table 5); the growth rate in most ponds was 0.4-0.6 mm/day (Figure 5). The regression between average daily growth rate and density of zooplankton was positive but non-significant ($r = 0.42$; $P < 0.20$). Based on unpublished observations of early growth rates of striped bass at study hatcheries in 1988, daily increments are relatively low

FIGURE 4. Average total length of striped bass harvested from each study pond as a function of number that were harvested. The open symbol represents a pond with a high number of fairy shrimp.

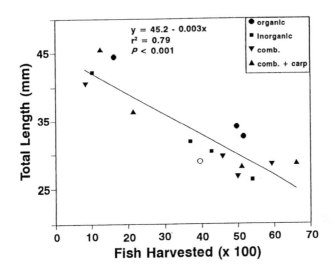

during this initial stage of growth and development. A growth rate ≥ 0.4 mm/day was observed in seven of eight ponds with average zooplankton densities of 10-20/L. Growth rate was only 0.27 mm/day in one pond with a zooplankton density of more than 40/L. This was the pond where dissolved oxygen was 4.6 mg/L on the morning of D5.

Maximum density of rotifers during this interval in any pond was 8.1/L; 26 of 32 samples had densities of <2.0/L. Larvae were dependent for food on recently hatched cladocerans and young copepods. The density of organisms of a size that could be consumed was less than the total crustacean zooplankton count.

From D5 to D8, the non-significant but positive relationship between average daily growth rate and fish harvest density ($r = 0.36$; $P < 0.20$) suggests there was no competition for food in this interval. There was also no significant correlation between the average zooplankton densities and number of fish harvested ($r = 0.19$; $P < 0.50$). Fish did not suppress numbers of zooplankton; higher densities of zooplankton did not enhance number of fish harvested.

FIGURE 5. Relationships of striped bass in the period from D5 to D8: average daily growth rate and average density of crustacean zooplankton (A); average daily growth rate and number of fish harvested (B); average density of crustacean zooplankton and number of fish harvested (C). Open symbols represent ponds with the highest numbers of fairy shrimp and clam shrimp.

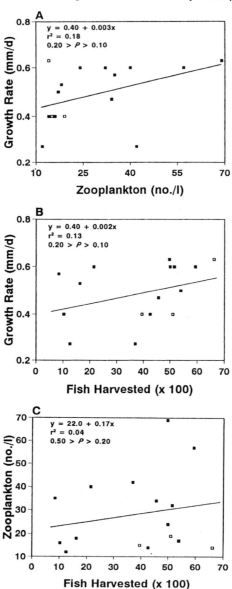

TABLE 5. Average growth rate (AGR) in mm/day and average crustacean zooplankton densities (no./L) for each 0.04-ha pond during four age intervals of striped bass production (D5-D8, D8-D19, D19-D25, and D25-harvest).

	D5-D8		D8-D19		D19-D25		D25-harvest	
	AGR	Zoopl	AGR	Zoopl	AGR	Zoopl	AGR	Zoopl
Organic	0.53	18	1.16	259	1.28	469	1.15	136
	0.60	32	1.03	108	0.72	39	0.60	8
	0.40	15	0.83	21	0.65	6	0.70	4
	0.63	69	0.85	260	0.97	166	0.84	22
Chemical	0.50	17	0.64	56	0.97	197	0.48	32
	0.40	14	0.83	270	0.90	196	0.65	11
	0.40	16	1.08	166	1.08	198	1.09	14
	0.27	42	0.95	372	1.10	294	0.56	16
Combination	0.47	34	0.95	72	0.92	38	0.43	15
	0.60	57	1.03	146	0.63	56	0.44	6
	0.60	24	0.85	89	0.82	46	0.33	7
	0.57	35	1.13	145	0.60	160	0.96	165
Combination +	0.63	14	0.91	60	0.60	29	0.53	10
common carp	0.40	19	0.85	38	0.75	26	0.45	14
	0.27	12	1.19	238	1.20	308	1.30	88
	0.60	40	1.13	107	0.98	239	0.84	30

From D8 to D19, the average density of zooplankton ranged from 21 to 372/L average growth rate ranged from 0.64 to 1.19 mm/day (Table 5). There was a positive but non-significant correlation ($r = 0.32$; $P < 0.50$) between average daily growth rate and zooplankton density (Figure 6). An average growth rate increment of ≥ 0.83 mm/day was observed in 15 out of 16 ponds. There was evidence of competition for food during this period, since the correlation between average daily growth rate and number at harvest was significant and negative ($r = -0.71$; $P < 0.005$). Based on the regression, the availability of food was sufficient to maintain an average growth rate of 0.83 mm/day at a fish density of 150,000/ha. Zooplankton densities of 100/L might be needed to sustain a satisfactory growth rate at a fish density of 250,000/ha. The numbers of zooplankton were not suppressed by the numbers of fish at harvest ($r = -0.36$; $P < 0.20$); lowest densities of zooplankton were in ponds with the highest numbers of fairy shrimp and clam shrimp.

From D19 to D25, the average density of zooplankton ranged from 6 to 469/L; average growth rate ranged from 0.60 to 1.28 mm/day (Table 5). There was a significant positive correlation ($r =$

FIGURE 6. Relationships of striped bass in the period from D8 to D19: average daily growth rate and average density of crustacean zooplankton (A); average daily growth rate and number of fish harvested (B); average density of crustacean zooplankton and number of fish harvested (C). Open symbols represent ponds with the highest numbers of fairy shrimp and clam shrimp.

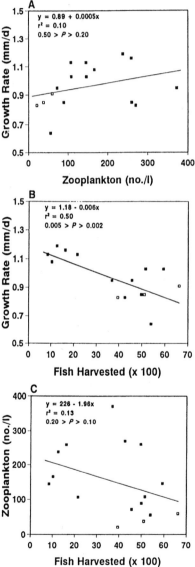

0.84; $P < 0.001$) between average growth rate and average zooplankton density (Figure 7). The regression intercepts the Y-axis at an increment <0.83 mm/day, which suggests that zooplankton were important to sustain satisfactory growth during this period. The significant negative correlation (r = -0.52; $P < 0.05$) between average daily growth rate and fish density at harvest indicates there was competition for food during this time interval. Based on the regressions, the availability of food was sufficient to maintain an average growth rate of 0.83 mm/day at a fish density of 125,000/ha. Zooplankton densities of 250/L might be needed to sustain a satisfactory average daily growth rate at a fish density of 250,000/ha. The numbers of zooplankton were probably suppressed by the numbers of fish (r = -0.65; $P < 0.01$).

In the final phase of production from D25 to harvest, average zooplankton densities ranged from 4 to 165/L; average growth rate ranged from 0.33 to 1.30 mm/day (Table 5). Only six populations exhibited an average increment ≥ 0.83 mm/day. The significant negative correlation (r = -0.85; $P < 0.001$) between growth rate and fish density at harvest indicated competition for food during this period (Figure 8). Based on the regression, the availability of food was sufficient to maintain an average growth rate of 0.83 mm/day at a fish density of 73,000/ha. The wide range of zooplankton density at a low fish harvest density suggested that other factors or predators were also influencing the density of zooplankton.

Less than satisfactory fish growth and production could be related to competition from predatory insects. Based on numbers observed with a light at night and collected in the fry seine, they were common in the ponds.

An early filling with an aggressive fertilization schedule prior to stocking was recommended by Geiger and Turner (1990) in order to maximize numbers of crustacean zooplankton when larvae were stocked (Geiger 1983); the recommended zooplankton density of 500/L is beyond the needs of larval striped bass in the initial phase of growth from D5 to D8. The need or benefit of high initial densities of zooplankton in striped bass ponds is undocumented. Geiger (1983) assumed that the quantity and quality of zooplankton was often too low, especially in the early culture period, because of low survival and evidence of cannibalism. Scarce food resources and

FIGURE 7. Relationships of striped bass in the period from D19 to D25: average daily growth rate and average density of crustacean zooplankton (A); average daily growth rate and number of fish harvested (B); average density of crustacean zooplankton and number of fish harvested (C). Open symbols represent ponds with the highest numbers of fairy shrimp and clam shrimp.

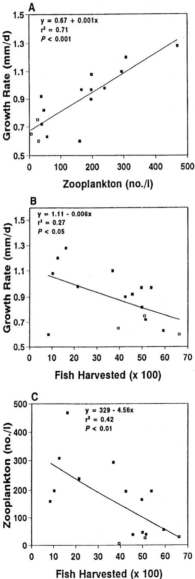

FIGURE 8. Relationships of striped bass in the period from D25 to harvest: average daily growth rate and average density of crustacean zooplankton (A); average daily growth rate and number of fish harvested (B); average density of crustacean zooplankton and number of fish harvested (C). Open symbols represent ponds with the highest numbers of fairy shrimp and clam shrimp.

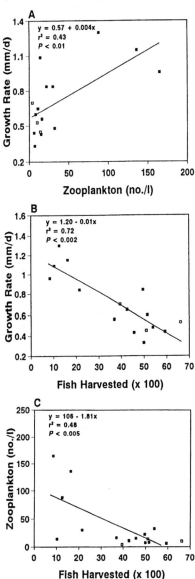

cannibalism are more likely to occur during the last half of the production period than during the initial stages. Striped bass larvae can survive for an extended period without food and do not exhibit a point of no return (Eldridge et al. 1981; Martin et al. 1985).

Counting zooplankton has been recommended as a routine hatchery practice (Geiger and Turner 1990). Hatchery managers should be familiar with the common taxa and expected population dynamics of zooplankton in their ponds. However, counting zooplankton is too labor intensive, and the time lag between sampling and counting is often too long to be used to make management decisions. Even if timely data were available, information on zooplankton density is of little or no value to "fine tune" pond fertilization practices, to decide how many larvae to stock, or to decide when to schedule harvests. The count on one day is information that can't be used with confidence to predict the count 2 or 3 days later. Since different taxa and life stages vary widely in size, a count obtained at one time or place may represent a different quantity of available food than the same count at a different time or place. There is no question that zooplankton are an important food source for striped bass larvae; however, they represent only a portion of what is available for fish. Given these uncertainties, defining good densities of zooplankton to support good fish production may be futile.

Ponds With Common Carp

The presence of common carp in four study ponds provided economic and ecological benefits. The primary benefit was effective control of *Chara*. When ponds without common carp were drained, a considerable number of fish were stranded in the vegetation (Table 4). The percentage of fish stranded was directly related to the length of fish ($r = 0.72$; $P < 0.005$) and plant biomass (Figure 9). These data suggest that if fish average 35 mm and there is a *Chara* biomass of 300-500 g/m^2, 5-25% of the fish may be stranded at harvest.

The presence of common carp increased turbidity and reduced Secchi disk transparency. The Secchi disk was visible on the bottom in ponds without common carp in 77% of measurements. In ponds with common carp, this was true only 6% of the time. Clam shrimp, as well as common carp, increase turbidity due to suspended sedi-

FIGURE 9. Relationships between percent of striped bass stranded in vegetation in ponds without common carp and average length of fish (A) and the adjusted percent stranded at an average length of 35 mm as a function of dry weight of plant biomass at harvest (B). The adjusted percent stranded was based on the slope of the regression in A (b = 1.25). The curve in Figure 9B was fit by eye.

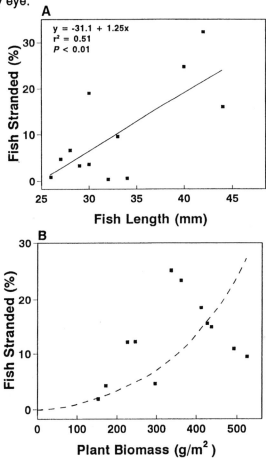

ments (McCraren et al. 1977); average Secchi disk transparency in the two ponds with common carp and clam shrimp was 67 and 104 cm; averages in the two ponds without clam shrimp were 106 and 108 cm. Suspended sediments are a site for bacterial development and a means to convert dissolved organic material to a food resource for zooplankton (Arruda et al. 1983). However, no enhance-

FIGURE 10. Relationship between average chlorophyll *a* concentration for the last two sample dates in each pond and the dry weight of plant biomass at harvest. The curve was fit by eye.

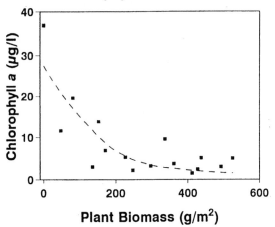

Plant Biomass (g/m²)

ment of zooplankton density was evident. Control of *Chara* may have made possible the higher concentrations of chlorophyll *a* during the final phase of the study (Figure 10).

Turbidity might be expected to influence feeding by larval striped bass. However, Chesney (1989) observed in laboratory experiments that turbidity due to 50, 100, and 150 ppm kaolin did not significantly affect growth. Breitburg (1988) measured feeding rates of striped bass larvae with suspended solids at 0, 75, 200, and 500 mg/L. He found that there was no effect of turbidity on consumption rate of *Daphnia* but that there was a reduction in consumption of copepods at 200 and 500 mg/L. The average NTU valves ($\approx$ mg/L) in ponds with common carp ranged from 2.5 to 9.8, well below values expected to influence feeding by striped bass larvae.

Ponds with common carp had an average pH equal to or less than the other treatments on 20 of 22 days after fertilization was initiated (Figure 1). The combination of common carp and an organic fertilizer should be a practical strategy in ponds with a history of problem plant growths and excessive pH; such problems are especially common in waters of low alkalinity.

Excessive pH is a serious problem when either high pH or a combination with un-ionized ammonia can cause toxicity to young larvae (Bergerhouse 1993). High pH may also influence qualitative aspects of algal production and ecological efficiency in the production of zooplankton. Classic studies by King (1970), which were corroborated by Shapiro (1973), documented that high pH and low free CO_2 promote the production of blue-green algae.

An abundance of algal filaments reduced the food gathering and reproductive output of *Daphnia* (Vaga et al. 1984). When *Daphnia* were fed *Aphanizomenon flos-aquae,* all of their lipid reserves were lost (Holm and Shapiro 1984). Growth, survival, and reproductive success of *Daphnia* are directly related to the quality and quantity of food resources (Richman 1958; Goulden et al. 1982a, 1982b; Tessier et al. 1983). Boom and bust population dynamics have been related to decreased food densities and depleted energy reserves in laboratory populations of *Daphnia* (Goulden and Hornig 1980). Similar dynamics may be likely in hatchery ponds. Cladocera cannot be stockpiled.

The average concentration of inorganic N was higher in the ponds with, than in ponds without, common carp (Table 1). This may have been due to nutrient cycling, lower photosynthetic uptake, or higher inflows of well water. Higher availability of P in ponds with common carp was expected (Lamarra 1975; Braband et al. 1990); however, concentrations of available P were similar in all ponds.

An additional benefit anticipated from the common carp was consumption of clam shrimp. High concentrations of these organisms can cause high turbidity, as well as plug screens, and can cause sorting problems when ponds are drained. The fecal material from all common carp examined contained remains of hundreds of clam shrimp. The second most common organism in fecal samples was large chironomid larvae. A few dytiscid larvae were also observed. No remains of zooplankton were detected. Control of clam shrimp was effective, since only a few were observed in one of the ponds when they were drained.

Cost and Other Considerations

There were no obvious treatment differences in the average dynamics of chlorophyll *a,* zooplankton, or production of striped bass. However, there was a considerable difference in fertilizer costs. The average cost with alfalfa meal was $155.75/ha compared to $14.75/ha for urea and phosphoric acid. The average cost of the combination of fertilizers in ponds with and without common carp were $67.75 and $71.25/ha, respectively. Less urea was added to the ponds with common carp (Table 6).

How much and what kind of fertilizer to apply to a given pond on a given day will always be an important management decision. The source and amounts of N and P to apply will have a direct influence on the cost of production, as well as indirect influences on water quality and the qualitative and quantative responses of aquatic plants. Optimum application amounts of N and P should vary with the age of fish and from pond to pond.

The study ponds developed a moderate phytoplankton bloom and concentration of chlorophyll *a* even though the initial concentration of P in the water supply was only 12 μg/L. However, available P at 28-38 μg/L each week did not provide adequate productivity for a high density of fish and fish food organisms. The greater need for P than that used by Culver et al. (1993) may be caused by the NFHTC water hardness of >200 mg/L as $CaCO_3$.

The 600 μg/L of available N each week was more than adequate to satisfy the need for primary production. The ratio of C:N:P in

TABLE 6. Average inputs (in grams) and cost per 0.04-ha pond of either nitrogen or phosphorus from well water, alfalfa meal, and chemicals (urea or phosphoric acid).

Nutrient/ treatment	Well water	Alfalfa meal (g)	Alfalfa meal ($)	Chemical (g)	Chemical ($)	Total (g)
Nitrogen						
Organic	1,022	748	6.22	0	0	1,770
Chemical	786	0	0	553	0.38	1,339
Combination	748	309	2.57	241	0.16	1,298
Combination + common carp	1,122	303	2.52	108	0.08	1,533
Phosphorus						
Organic	10.7	49.9	6.22	3.4	0.01	64.0
Chemical	8.2	0	0	61.7	0.21	69.9
Combination	7.8	26.9	2.57	26.9	0.12	61.6
Combination + common carp	11.8	26.2	2.52	26.2	0.11	64.2

phytoplankton is approximately 40:7:1 (Round 1973; Vallentyne 1974; Wetzel 1983). However, a ratio of N:P > 7:1 may be beneficial if the rate of regeneration of P is higher than that of N. A lower N:P ratio may be in order if internal cycling of TAN is faster than that of P.

The optimum ratio and concentrations of available N and P that should be achieved in fertilized hatchery ponds is likely related to hatchery water chemistry and the species of fish in production. Further research is needed to define optimality and to achieve better efficiency and success.

ACKNOWLEDGMENTS

This study could not have been conducted without the analyses of NO_3^--N, SRP, and chlorophyll *a* conducted at Southwest Texas State University by A. Groeger. Staff at the NFHTC were also essential for assistance during the study and for preparation and editorial review of this paper. C. Berkhouse provided invaluable assistance and support in the preparation of figures and statistical analyses.

REFERENCES

Albrect, A. B. 1964. Some observations on factors associated with survival of striped bass eggs and larvae. California Fish and Game 50:100-113.

Anderson, R. O. 1993a. New approaches to management of fertilized hatchery ponds. Journal of Applied Aquaculture 2(3/4):1-8.

Anderson, R. O. 1993b. Apparent problems and potential solutions for production of fingerling striped bass, *Morone saxatilis.* Journal of Applied Aquaculture 2(3/4):101-118.

Arruda, J. A., G. R. Marzolf, and R. T. Faulk. 1983. The role of suspended sediments in the nutrition of zooplankton in turbid reservoirs. Ecology 64:1225-1235.

Bergerhouse, D. L. 1989. Lethal Effects of Elevated pH and Ammonia on Early Life Stages of Several Sportfish Species. Doctoral dissertation, Southern Illinois University Carbondale. Illinois.

Bergerhouse, D. L. 1993. Lethal effects of elevated pH and ammonia on early life stages of hybrid striped bass. Journal of Applied Aquaculture 2(3/4):81-100.

Braband, A., B. A. Faafeng, and J. P. M. Nilssen. 1990. Relative importance of phosphorus supply to phytoplankton production: Fish excretion versus external loading. Canadian Journal of Fisheries and Aquatic Sciences 47:364-372.

Breitburg, D. L. 1988. Effects of turbidity on prey consumption of striped bass larvae. Transactions of the American Fisheries Society 117:72-77.

Brewer, D. L., and R. A. Rees. 1990. Pond culture of phase I striped bass fingerlings. Pages 99-120 *in* R. M. Harrell, J. H. Kerby, and R. V. Minton, eds. Culture and Propagation of Striped Bass and its Hybrids. Striped Bass Committee, Southern Division, American Fisheries Society, Bethesda, Maryland.

Bulak, J. S., and R. C. Heidinger. 1980. Developmental anatomy and inflation of the gas bladder in striped bass, *Morone saxatilis.* Fishery Bulletin 77:1000-1003.

Burleigh, J. G., R. W. Katayama, and N. Elkassabany. 1993. Impact of predation by backswimmers in golden shiner, *Notemigonus crysoleucas,* production ponds. Journal of Applied Aquaculture 2(3/4):243-256.

Burnison, B. K. 1980. Modified dimethyl sulfoxide (DMSO) extraction for chlorophyll analysis of phytoplankton. Canadian Journal of Fisheries and Aquatic Sciences 37:729-733.

Chesney, E. J., Jr. 1989. Estimating the food requirements of striped bass larvae, *Morone saxatilis*: Effects of light, turbidity and turbulence. Marine Ecology Progress Series 53:191-200.

Culver, D. A., S. Madon, and J. Qin. 1993. Percid pond production techniques: Timing, enrichment, and stocking density manipulation. Journal of Applied Aquaculture 2(3/4):9-31.

Doroshev, S. I. 1970. Biological features of the eggs, larvae and young of striped bass [*Roccus saxatilis* (Walbaum)] in connection with the problem of its acclimatization in the USSR. Journal of Ichthyology 10:235-248.

Doroshov, S. I., and J. W. Cornacchia. 1979. Initial swimbladder inflation in the larvae of *Tilapia mossambica* (Peters) and *Morone saxatilis* (Walbaum). Aquaculture 16:57-66.

Eldridge, M. B., J. A. Whipple, D. Eng, M. J. Bowers, and B. M. Jarvis. 1981. Effects of food and feeding factors on laboratory-reared striped bass larvae. Transactions of the American Fisheries Society 110:111-120.

Geiger, J. G. 1983. A review of pond zooplankton production and fertilization for the culture of larval and fingerling striped bass. Aquaculture 35:353-369.

Geiger, J. G., and C. J. Turner. 1990. Pond fertilization and zooplankton management techniques for production of fingerling striped bass and hybrid striped bass. Pages 79-98 *in* R. M. Harrell, J. H. Kerby, and R. V. Minton, eds. Culture and Propagation of Striped Bass and its Hybrids. Striped Bass Committee, Southern Division, American Fisheries Society, Bethesda, Maryland.

Goulden, C. E., and L. L. Hornig. 1980. Population oscillations and energy reserves in planktonic cladocera and their consequences to competition. Proceedings of the National Academy of Science 77:1716-1720.

Goulden, C. E., L. L. Henry, and A. J. Tessier. 1982a. Body size, energy reserves, and competitive ability in three species of Cladocera. Ecology 63:1780-1789.

Goulden, C. E., R. M. Comotto, J. A. Hendrickson, Jr., L. L. Hornig, and K. L. Johnson. 1982b. Procedures and Recommendations for the Culture and Use of *Daphnia* in Bioassay Studies. Technical Publication 766, American Society for Testing and Materials, Philadelphia, Pennsylvania.

Graves, K. G., and J. C. Morrow. 1988a. Tube sampler for zooplankton. Progressive Fish-Culturist 50:182-183.

Graves, K. G., and J. C. Morrow. 1988b. Method for harvesting large quantities of zooplankton from hatchery ponds. Progressive Fish-Culturist 50:184-186.

Haney, J. F., and D. J. Hall. 1973. Sugar-coated *Daphnia*: A preservation technique for Cladocera. Limnology and Oceanography 18:331-333.

Holm, N. P., and J. Shapiro. 1984. An examination of lipid reserves and the nutritional status of *Daphnia pulex* fed *Aphanizomenon flos-aquae*. Limnology and Oceanography 29:1137-1140.

King, D. L. 1970. The role of carbon in eutrophication. Journal of the Water Pollution Control Federation 42:2035-2051.

Lamarra, V. A. 1975. Digestive activities of carp as a major contributor to the nutrient loading in lakes. International Association of Theoretical and Applied Limnology 19:2461-2468.

McCraren, J. P., J. L. Millard, and A. M. Woolven. 1977. Masoten (Dylox) as a control for clam shrimp in hatchery production ponds. Proceedings of the Southeastern Association of Fish and Wildlife Agencies 31:329-331.

Martin, F. D., D. A. Wright, J. C. Means, and E. M. Setzler-Hamilton. 1985. Importance of food supply to nutritional state of larval striped bass in the Potomic River estuary. Transactions of the American Fisheries Society 114:137-145.

Piper, R. G., I. B. McElwain, L. E. Orme, J. P. McCraren, L. G. Fowler, and J. R. Leonard. 1982. Fish Hatchery Management. United States Department of the Interior, Fish and Wildlife Service, Washington, D. C.

Richman, S. 1958. The transformation of energy by *Daphnia pulex*. Ecological Monographs 28:273-291.

Round, F. E. 1973. The Biology of the Algae, 2nd ed. St. Martins Press, New York, New York.

Shapiro, J. 1973. Blue-green algae: Why they become dominant. Science 179:382-384.

Tessier, A. J., L. L. Henry, C. E. Goulden, and M. W. Durando. 1983. Starvation in Daphnia: Energy reserves and reproductive allocation. Limnology and Oceanography 28:667-676.

Turner, C. J. 1988. Detection of stocking mortality in striped bass and hybrid striped bass culture ponds. Progressive Fish-Culturist 50:124-126.

Vaga, R. M., D. A. Culver, and C. S. Munch. 1984. The fecundity ratios of *Daphnia* and *Bosmina* as a function of inedible algal standing crop. International Association of Theoretical and Applied Limnology 22:3072-3075.

Vallentyne, J. R. 1974. The Algal Bowl; Lakes and Man. Special Publication 22, Department of the Environment, Fisheries and Marine Science, Ottawa, Canada.

Wetzel, R. G. 1983. Limnology, 2nd ed. Saunders College Publishing, New York, New York.

Zar, J. H. 1984. Biostatistical Analysis. Prentice-Hall, Inc., Englewood Cliffs, New Jersey.

Effects of Pond Volume Manipulation on Production of Fingerling Largemouth Bass, *Micropterus salmoides*

Aaron Barkoh
Richard O. Anderson
Charles F. Rabeni

ABSTRACT. Hatchery ponds were manipulated by timing of filling to increase the efficiency of production of fingerling largemouth bass, *Micropterus salmoides*. The effects on pond productivity of fertilizing with chicken manure versus plant meals (consisting of equal parts by weight of cottonseed meal, alfalfa meal, and wheat shorts) were also examined. When ponds were filled in stages, more favorable water temperatures for largemouth bass were achieved, and desirable dissolved oxygen concentrations were maintained. Chlorophyll *a* levels and plankton community respiration rates were higher in ponds that were fertilized with the plant meals than in ponds

Aaron Barkoh, Department of Fisheries and Wildlife, School of Natural Resources, University of Missouri, Columbia, MO 65211, USA. Correspondence may be addressed to Texas Parks and Wildlife Department, Rt. 1, Electra, TX 76360, USA.

Richard O. Anderson and Charles F. Rabeni, Missouri Cooperative Fish and Wildlife Research Unit, School of Natural Resources, University of Missouri, Columbia, MO 65211, USA.

Correspondence for Richard O. Anderson may be addressed to 3618 Elms Court, Missouri City, TX 77459, USA.

[Haworth co-indexing entry note]: "Effects of Pond Volume Manipulation on Production of Fingerling Largemouth Bass, *Micropterus salmoides*." Barkoh, Aaron, Richard O. Anderson, and Charles F. Rabeni. Co-published simultaneously in the *Journal of Applied Aquaculture*, (The Haworth Press, Inc.) Vol. 2, No. 3/4, 1993, pp. 151-170; and: *Strategies and Tactics for Management of Fertilized Hatchery Ponds* (ed: Richard O. Anderson and Douglas Tave) The Haworth Press, Inc., 1993, pp. 151-170. Multiple copies of this article/chapter may be purchased from The Haworth Document Delivery Center [1-800-3-HAWORTH; 9:00 a.m. - 5:00 p.m. (EST)].

151

fertilized with chicken manure (25% dry matter). Secchi disk visibility was relatively greater in chicken manure ponds than in ponds fertilized with the plant meals; however all differences were not significant. Zooplankton densities were comparable among treatments. Filling ponds in stages had no significant effects on chlorophyll a, plankton community respiration rate, Secchi disk visibility, or zooplankton abundance and population dynamics. Daily production of largemouth bass was greater in ponds filled in stages and fertilized with plant meals than in other treatment ponds. Yield was highest in stage-filled ponds fertilized with plant meals and lowest in full ponds fertilized with chicken manure. Filling ponds in stages reduced the fertilizer cost per kilogram of largemouth bass by 50% in ponds fertilized with plant meals and by 65% in manure-fertilized ponds.

INTRODUCTION

Culture of larval fish in earthen ponds requires the establishment and production of zooplankton and desirable water quality conditions. Fertilization regimens that include frequent applications of organic and inorganic fertilizers provide the best conditions for fish production (Hepher 1963; Sobue et al. 1977; Wohlfarth and Schroeder 1979). However, fertilizing ponds with organic fertilizers can be expensive (Piper et al. 1982); at some hatcheries, increasing fertilizer costs have reduced fertilizer application rates (Farquhar 1984; Geiger and Parker 1985). Additionally, fertilized ponds have rarely sustained high concentrations of zooplankton throughout the period that they are the preferred food organisms of larval fish. Pennak (1978) identified several factors, including temperature, pH, and excretory products, that might influence zooplankton populations. In laboratory cultures, zooplankton production and densities have been maintained by periodic dilution of culture media (Ivleva 1973). This concept has not been investigated in fish hatchery ponds.

Water temperature has considerable regulatory influence on pond productivity. Low temperatures delay phytoplankton response to fertilization (Noriega-Curtis 1979; Doyle and Boyd 1984) and slow down zooplankton production (Welch 1980). Conversely, high temperatures can be lethal to zooplankton (Welch 1980) or can create water quality conditions that favor blue-green algae over desirable green algae (McQueen and Lean 1987). Fish feeding activities and growth rates are also temperature-dependent (Dupree and Huner 1984), yet hatchery managers usually do not manage for water temperatures in fingerling production ponds. Recently, the need to purposefully seek favorable temperatures for fish has been emphasized (Coutant 1987).

Pond management techniques that can reduce operating costs and promote rapid growth of cultured fish would increase net economic return (Amir 1987). This study was conducted to investigate the effects that filling ponds in stages may have on water quality, zooplankton densities, fish production, and cost of pond fertilization. In addition, the efficacy of either chicken manure or plant meals in promoting pond productivity was examined. A few studies (e.g., Geiger et al. 1985) have shown that greater fish yields were produced in ponds fertilized with vegetable meals than in ponds that were fertilized with chicken manure. Those studies were conducted without pond volume manipulation. Largemouth bass, *Micropterus salmoides*, was used for this study because it is an important game fish that grows best at temperatures above 20°C and because fry were available locally. An important hypothesis of this study is that desirable temperatures, dissolved oxygen concentrations, and productivity are attainable with pond volume manipulation.

MATERIALS AND METHODS

Pond Preparation and Fertilization Technique

The study was conducted in experimental ponds that are owned by the Missouri Department of Conservation and that are located near Columbia, Missouri. Eight 0.2-ha ponds were drained in February, 1984 and allowed to dry.

Filling of ponds was started on May 4 or 8, and largemouth bass fry were stocked 9 or 13 days after pond filling started. All water came from Little Dixie Lake, Missouri and was strained through a saran bag (mesh size = 1 mm^2) to exclude wild fish and predatory invertebrates. Four ponds were completely filled and four were one-third filled on May 9; these ponds were designated as "full" and "stage-filled," respectively. The stage-filled ponds were two-thirds filled during the second week (May 27) after fry were stocked and were completely filled during the fourth week (June 11).

Pond fertilization began May 7-11. Ponds were fertilized with liquid 10-34-0 (N-P_2O_5-K_2O) and either chicken manure or a combination of equal parts (by weight) of cottonseed meal, alfalfa meal, and wheat shorts (plant meals). The liquid inorganic fertilizer was applied at a rate and ratio of N:P = 470:750 μg/L/week in one or two applications. The liquid fertilizer was diluted with pond water before it was broadcast onto the pond surface. Moist chicken manure (25% dry matter) was applied at a rate and ratio of N:P = 219:67 μg/L/week in either one or two applications in the first week and at N:P = 158:49 μg/L/week in the second week. The rate was reduced to N:P = 105:32 μg/L/week in subsequent weeks. The manure was mixed in pond water, and the suspended slurry was distributed along the edge of the pond. The plant meals were added to ponds at a weekly rate of N:P = 2,442:149 μg/L/week in one or two applications for the first two weeks; fertilization rate each week thereafter was N:P = 1,221:75 μg/L/week. The plant meals were broadcast dry onto the water surface.

Water Quality

Water quality was monitored weekly in all ponds, and all samples were collected between 0830 and 1130 hours. Measurements were made at the drain box area. Water temperature and dissolved oxygen concentration profiles were taken at 0.5-m depth intervals using a YSI[1] Model 54A oxygen meter. Simultaneously, Secchi disk visibility was measured. An integrated water sample was taken to a depth of 1.2-1.5 m with a 3-cm-diameter PVC pipe in each pond. The sample was agitated so that it was saturated with dissolved oxygen, and it was then siphoned into two 300-ml BOD bottles to determine plankton community respiration rate. The dark bottles were incubated for 24 hours at room temperature. Oxygen depletion in these bottles was determined by the Winkler method (APHA et al. 1980). Chlorophyll *a* was also determined from the integrated water samples. Chlorophyll *a* concentration was measured using

---

1. Use of trade name does not imply endorsement.

acetone-DMSO (dimethyl sulfoxide) extraction and the fluorometry procedure described by Knowlton (1984).

Zooplankton Populations

Zooplankton samples were collected weekly by taking two vertical tows with a 21-cm-diameter Wisconsin plankton net (80-μ mesh) at the drain box area. Plankton tows were taken from a depth of 1.2 m when ponds were not full and from a 1.8-m depth when ponds were full. Zooplankters were killed quickly with commercial strength 37% formalin and were then preserved in 4% buffered formalin. Zooplankters were identified to genus, using the taxonomic key in Pennak (1978) and were enumerated into nine categories–*Daphnia, Ceriodanhnia*, other cladocerans, Cladocera, *Cyclops*, nauplii, Crustacea, rotifers, and total zooplankton. Densities were expressed as number of organisms/L of pond water.

Fish Production

Largemouth bass fry (390 fish/g; 7 mm average total length) were stocked at rates of 88,462-94,392 fry/ha 9 or 13 days after pond filling started. Fingerling largemouth bass were harvested 28-33 days poststocking. All ponds were harvested by draining the water and dipping the fish from the drain box. All fish recovered from each pond were weighed, and the number of fish was estimated from weighed and counted samples. These data were used to calculate yield, percent survival, and daily production. Mean total length at harvest was determined from a sample of 150-200 fish from each pond.

Data Analysis

Data were analyzed for treatment and sampling date effects by split-plot analysis (Gill and Hafs 1971), using the General Linear Models procedure of the Statistical Analysis System (SAS Institute, Inc. 1985). When there was a significant treatment-date interaction, differences between treatment means by date were tested with LSD (least significant difference). Overall treatment means were compared with Duncan's multiple-range test (Snedecor and Cochran 1980).

Temperature and dissolved oxygen profile values for each sampling date were averaged over depth and designated as water column temperature and water column dissolved oxygen concentration, respectively. Water column values, surface values 0.5 m below the air-water interface, and bottom values 0.5 m above the bottom mud were used for analysis. Chlorophyll *a* concentration, plankton community respiration rate, and Secchi disk visibility data were $\log_{10}$ transformed before statistical analysis to reduce heterogeneity of variances. Zooplankton categories *(Daphnia, Ceriodaphnia,* other cladocerans, total Cladocera, *Cyclops,* nauplii, Crustacea, rotifers, and total zooplankton) were analyzed separately. All zooplankton data were $\log_{10}(X + 1)$ transformed before statistical analysis. Differences were considered significant at $P \leq 0.05$.

RESULTS AND DISCUSSION

Water Quality

Mean bottom and mean water column temperatures were significantly higher in stage-filled ponds than in full ponds (Table 1; Figure 1). The low temperature which was observed on day 24 for stage-filled ponds was due to the addition of relatively cool water from Little Dixie Lake. The high surface area:volume ratio of stage-filled ponds partly explains the observed high temperatures in these ponds during the early part of the study. The small volume of water in these ponds, during the beginning, was quickly heated by the sun. Filling ponds in stages thus has the potential to attain higher temperatures to promote higher growth rates of cultured fish. During the first 18 days, the temperature in stage-filled ponds was 22-24°C, which potentially could promote more rapid feeding and growth of largemouth bass than the range of 19-20.1°C that existed in the full ponds (Coutant 1975).

Dissolved oxygen concentrations were higher in stage-filled ponds than in full ponds, but the differences were significant only for bottom dissolved oxygen (Table 1). Temporally, water column

TABLE 1. Mean ± SE water quality variables in largemouth bass ponds subjected to two filling and two fertilization regimens. Values in a row bearing the same letter are not significantly different (*P* > 0.05). Plant meals is a combination of cottonseed meal, alfalfa meal, and wheat shorts.

Water quality variable	Chicken manure		Plant meals	
	Stage-filled	Full	Stage-filled	Full
Bottom temperature (°C)	19.7±0.7a	15.8±0.7b	20.1±1.1a	15.2±0.5b
Water column temperature (°C)	22.1±0.9a	20.6±0.8b	21.5±1.1a	20.1±0.7b
Bottom DO (mg/l)	4.6±0.9a	1.2±0.5b	1.8±0.8b	0.4±0.1c
Water column DO (mg/l)	7.1±0.5a	5.7±0.4a	6.4±0.5a	4.7±0.8a
Chlorophyll a (ug/l)	13.6±2.8a	16.0±2.4a	35.8±8.5b	46.7±11.3b
Plankton community respiration (mg/l)	1.7±0.3a	1.9±0.3a	3.4±0.4b	3.6±0.5b
Secchi visibility (cm)	58.0±5.3a	78.0±6.4b	41.0±3.9a	49.0±4.5a

FIGURE 1. Weekly mean water temperatures of largemouth bass culture ponds that were either filled initially (full) or filled in stages (stage-filled) and fertilized with either a combination of plant meals or with chicken manure and 10-34-0 (N-P_2O_5-K_2O) fertilizer.

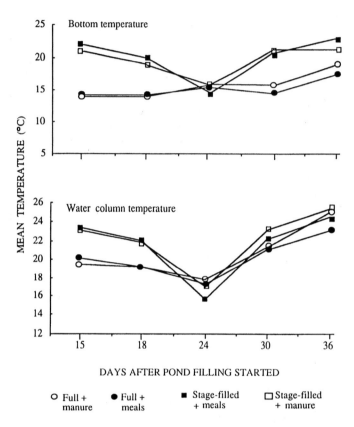

dissolved oxygen was significantly higher on days 15 and 18 in stage-filled ponds than in full ponds that received the same fertilizer treatments (Figure 2). Oxygenation by photosynthesis was probably reduced in the deeper water of full ponds; in stage-filled ponds all or a higher proportion of the water volume was illuminated and oxygenated by photosynthesis during daylight hours when these ponds were not full. Intermittent filling of ponds added cooler aerated water, which probably went to the bottom. Also, wind-induced water circulation may have been more effective in mixing oxygen-

FIGURE 2. Weekly mean dissolved oxygen in largemouth bass culture ponds that were either filled initially (full) or filled in stages (stage-filled) and fertilized with either a combination of plant meals or with chicken manure and 10-34-0 (N-P$_2$O$_5$-K$_2$O) fertilizer.

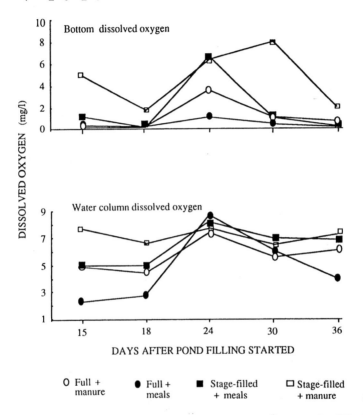

O Full +   ● Full +   ■ Stage-filled   □ Stage-filled
   manure   meals   + meals   + manure

rich upper layers of water with bottom water in ponds filled in stages, resulting in higher levels of oxygen at the bottom of these ponds.

All three productivity bioindicators were statistically different between fertilizer treatments. Mean chlorophyll *a* values were 35.8 and 46.7 μg/L in ponds fertilized with plant meals but were less than one-half of these levels in ponds that were fertilized with chicken manure (Table 1). Mean plankton community respiration rates were significantly higher in ponds fertilized with the plant meals. Observed mean Secchi disk visibilities were less in ponds fertilized with the plant meals; however all differences were not significant. Geiger et al. (1985) obtained similar results in ponds with striped bass, *Morone saxatilis*, that were fertilized with either cottonseed meal or combined chicken manure and chicken litter without pond volume manipulation. The pond-filling strategy used in this study did not significantly change the effects of these fertilizing regimens.

Chlorophyll *a* concentrations, Secchi disk visibilities, and plankton community respiration rates during the study are presented in Figure 3. A consistent pattern of better phytoplankton standing crops was observed in plant meals-fertilized ponds than in ponds that were fertilized with chicken manure. Geiger et al. (1985) attributed such an observation to the higher nitrogen content of cottonseed meal compared with that of chicken manure plus chicken litter. Because the nitrogen content of the chicken manure used in this study was far less than that of the plant meals, these results suggest that the amount of chicken manure needed to be increased in order to increase plankton production.

Zooplankton Populations

Zooplankton populations were dominated by *Daphnia*, cyclopoid copepods, and rotifers. The "other cladocerans" category consisted of *Bosmina, Diaphanosoma*, and chydorids, in descending order of abundance. Mean densities ranged from 717 to 923 organisms/L for total zooplankton, 284 to 449 organisms/L for *Daphnia*, and 80 to 183 organisms/L for *Cyclops*. The differences in densities of the dominant zooplankters were not significant (Table 2), which suggests that zooplankton production and loss rates were comparable in all treatments despite the relatively high phytoplankton standing crops in the plant meals-fertilized ponds. Farquhar (1984) also did not find differences in the densities of *Ceriodaphnia*, copepods, rotifers, or total zooplankton in ponds with Florida largemouth bass, *M. salmoides floridanus*, that were subjected to different fertilization techniques.

Temporal fluctuations in densities of predominant and important zooplankters are presented in Figure 4. On day 18, the population level of *Daphnia* was significantly higher in stage-filled ponds than in full ponds fertilized with plant meals. *Daphnia* densities were comparable in treatment ponds on the remaining sampling days.

FIGURE 3. Mean chlorophyll *a* concentrations, Secchi disk visibilities, and plankton community respiration rates in largemouth bass culture ponds that were either filled initially (full) or filled in stages (stage-filled) and fertilized with either a combination of plant meals or with chicken manure and 10-34-0 $(N-P_2O_5-K_2O)$ fertilizer.

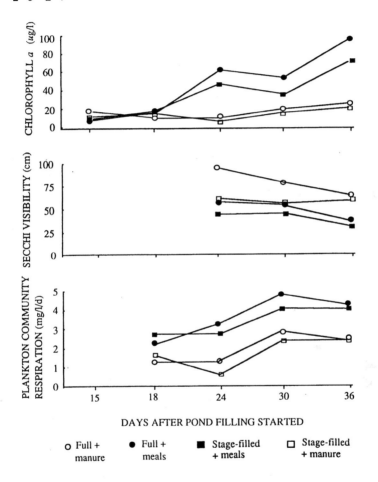

DAYS AFTER POND FILLING STARTED

o Full +   • Full +   ■ Stage-filled   □ Stage-filled
  manure   meals   + meals   + manure

Densities of *Ceriodaphnia* peaked after *Daphnia* populations declined. The density of *Ceriodaphnia* fluctuated in a similar pattern in all ponds and did not differ significantly among treatments on any sampling date.

Temporal variation of *Cyclops* densities differed considerably

TABLE 2. Mean ± SE numbers of zooplankton per liter in largemouth bass ponds subjected to two filling and two fertilization regimens. Plant meals is a combination of cottonseed meal, alfalfa meal, and wheat shorts. There were no significant differences among treatments ($P > 0.05$).

Organism	Chicken manure		Plant meals	
	Stage-filled	Full	Stage-filled	Full
Daphnia	449±136	424±143	490±208	284±48
Ceriodaphnia	14±9	36±15	16±9	29±20
Other cladocerans*	62±22	216±55	155±47	181±58
Total Cladocera	525±141	676±142	661±230	494±103
Cyclops spp.	183±46	114±20	80±10	137±22
Nauplii	25±13	41±11	3±1	18±5
Crustacea	708±165	790±149	741±228	631±122
Rotifers	128±49	66±18	59±20	56±16
Total zooplankton	865±210	923±177	813±230	717±131

*Includes Bosmina, Diaphanosoma, and the chydoridae.

FIGURE 4. Mean numbers per liter of *Daphnia, Ceriodaphnia,* and *Cyclops* in largemouth bass culture ponds that were either filled initially (full) or filled in stages (stage -filled) and fertilized with either a combination of plant meals or with chicken manure and 10-34-0 (N-P$_2$O$_5$-K$_2$O) fertilizer.

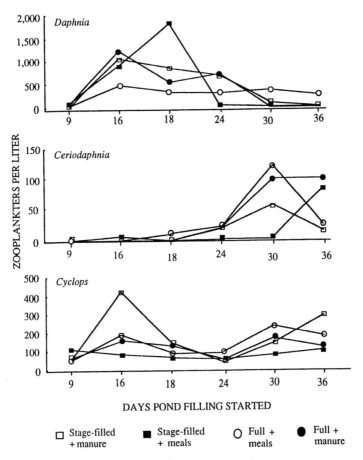

among treatments. On days 16 and 36, *Cyclops* densities were higher in stage-filled ponds receiving chicken manure than in stage-filled ponds fertilized with plant meals. Densities of *Cyclops* did not differ among treatments during the remaining sampling days. The differences in densities of dominant zooplankters among treatments over time did not follow a consistent pattern which could be attributed to type of fertilizer or method of pond filling. Inconsistent patterns of zooplankton population dynamics in managed fish ponds are commonplace (Geiger 1983). Filling ponds in stages had no effect on temporal variation of zooplankton densities.

Largemouth Bass Production

Mean survival rate at harvest ranged from 49 to 61% (Table 3), but the differences among treatments were not significant. Mean length at harvest was significantly smaller in full ponds that were fertilized with chicken manure than in other treatment ponds. Mean daily production ranged from 2.43 to 4.32 kg/ha/day and was significantly higher in ponds that were filled in stages and fertilized with plant meals. Also, mean yield was significantly higher in these ponds.

Geiger et al. (1985) produced more fingerling striped bass/ha in ponds that were fertilized with cottonseed meal than in those fertilized with combined chicken manure and chicken litter. They attributed this observation to high crustacean zooplankton densities in the ponds that received cottonseed meal. Because the zooplankton food base of the ponds of this study did not differ significantly among treatments, it is logical to assume that the observed differences in yield and daily production could be attributed, at least partially, to differences in heterotrophic production available to the fish (Noriega-Curtis 1979). Food habit studies have established that immature insects, such as chironomids and chaoborins, are important food items of postlarval largemouth bass (Hodson and Strawn 1968; Parmley et al. 1986).

High temperatures of 20-27°C (Coutant 1975) and high dissolved oxygen concentrations of ≥ 6 mg/L (Bulkley 1975) would promote largemouth bass production. During the first 18 days of the study, water column temperatures in stage-filled ponds averaged 22-24°C, while those of full ponds averaged 19-20.1°C. Also, dissolved oxygen concentrations averaged 5.0-8.0 mg/L in stage-filled ponds and 2.3-5.0 mg/L in full ponds. It is apparent that the stage filling protocol plus plant meals fertilization provided the best conditions for largemouth bass production.

Fertilizer Cost Comparison

A total of 406.8 kg of plant meals was used to fertilize each of two stage-filled ponds, and the quantity applied to each of two full

TABLE 3. Production summary for fingerling largemouth bass harvested from 0.2-ha ponds subjected to two filling and two fertilization regimens. Plant meals is a combination of cottonseed meal, alfalfa meal, and wheat shorts. Manure is chicken manure (25% dry weight). Mean ± SE values in each column bearing the same letter are not significantly different ($P > 0.05$).

Treatment	Yield (kg/ha)	Production (kg/ha/day)	Survival (%)	Mean length (mm)
Stage filled--plant meals	133.82±2.29a	4.32±0.08a	61±8.5a	55.4±1.7a
Full--plant meals	99.00±13.0b	3.41±0.45ab	49±5.0a	53.0±0.2a
Stage filled--manure	84.26±7.59bc	2.55±0.23b	56±0.5a	51.0±0.2a
Full--manure	67.81±0.68c	2.43±0.03b	55±7a	46.0±1.7b

ponds was 613.2 kg. Corresponding quantities of liquid 10-34-0 fertilizer were 21.2 kg and 23.0 kg, respectively. Chicken manure totalled 479.2 kg in each of two stage-filled ponds and 720 kg in the full ponds. Liquid 10-34-0 fertilizer in manure-fertilized ponds totalled 21.2 kg in each stage-filled pond and 24 kg in each full pond.

The costs for 45.4 kg of cottonseed meal, alfalfa meal, and wheat shorts were US$16.40, US$11.00, and US$8.20, respectively. The cost per 45.4 kg of combined plant meals was US$11.87. Chicken manure was estimated at a cost of US$0.44 per 100 kg, while liquid 10-34-0 fertilizer was purchased at a cost of US$0.29/kg (US$0.36/L). The total cost of fertilizer applied to ponds that were stage-filled and fertilized with plant meals was US$562.95/ha (Table 4); corresponding cost in full ponds was $835.67/ha. In chicken manure ponds, the total cost of fertilizer was US$41.25/ha for stage-filled ponds and US$50.65/ha for full ponds.

The cost of producing largemouth bass fingerlings in ponds fertilized with plant meals, in terms of cost of fertilizer per kilogram of fish, averaged US$4.21 for stage-filled ponds, compared with US$8.44 for full ponds; the corresponding costs for ponds fertilized with chicken manure were US$0.49 and US$0.75.

A potential saving of fertilizer costs was realized by filling ponds in stages. Less fertilizer was used in stage-filled ponds with either no loss in fish production or with higher yield from ponds that were fertilized with plant meals. Filling ponds in stages reduced the fertilizer cost per kilogram of fish produced by 50% in ponds fertilized with plant meals and by 65% in ponds fertilized with chicken manure.

An important constraint to pond fertilization with organic materials has been cost (Piper et al. 1982). The cost per unit N-P-K nutrient is higher for organic than for inorganic fertilization. However, because organic fertilizers produce greater yields, culturists are increasingly using these fertilizers in fish production ponds.

This study addresses one aspect of cost of pond fertilization–the quantity of fertilizer applied to ponds. The cost of pond fertilization is rarely reported by fish culturists. A survey of hatcheries in Missouri revealed that costs of fertilization vary considerably. The 1987 figures ranged from $1.97 to $8.71/kg of fish produced for paddlefish, *Polyodon spathula*, and walleye, *Stizostedion vitreum*, and

TABLE 4. Fertilizer amounts and mean (ranges in parentheses) costs of producing largemouth bass in 0.2-ha ponds subjected to two filling regimens: initially filled completely (full) or stage-filled. Plant meals is a combination of cottonseed meal, alfalfa meal, and wheat shorts; inorganic is 10-34-0 (N-P_2O_5-K_2O) fertilizer.

| | Plant meals | | Chicken manure | |
	Stage-filled	Full	Stage-filled	Full
Fertilizer (kg) per pond				
Organic	406.8	613.2	479.2	720.0
Inorganic	21.2	23.0	21.2	24.0
Cost ($/ha)				
Organic	532.25	802.32	10.55	15.85
Inorganic	30.70	33.35	30.70	34.80
Total	562.95	835.67	41.25	50.65
Cost of producing fish				
$/kg fish	4.21 (4.13-4.29)	8.44 (7.46-9.72)	0.49 (0.45-0.54)	0.75 (0.74-0.75)

from $2.12 to $4.73/kg of largemouth bass (5.1 cm total length). Hamilton (E. J. Hamilton, Blind Pony Hatchery and Wildlife Area, Sweet Springs, Missouri, pers. comm.) produced 5-cm walleyes at $8.71/kg and 25-cm paddlefish at $1. 97/kg with the same fertilization program over different culture periods. Bowling et al. (1984) produced 5-cm Florida largemouth bass at an average cost of $3.40/kg. All of these cost figures were associated with fertilizing ponds that were initially filled completely. The differences in the costs of producing 5-cm largemouth bass emanate from differences in fertilization rate and purchase price of fertilizer. These reasons also account for the differences between the cost figures in this study and the 1987 Missouri figures or that from Bowling et al. (1984).

The results of this study demonstrated that by filling ponds in stages, hatchery managers could reduce total quantities of fertilizers applied to ponds while maintaining existing fertilization rates. Consequently, it should be possible to reduce the cost of fertilization and simultaneously increase fingerling fish production.

ACKNOWLEDGMENTS

The authors wish to thank Steve Olson for his technical assistance during the study and Gene McCarty and Barbara Gregg for reviewing the manuscript. This is a contribution from the Missouri Cooperative Fish and Wildlife Research Unit (U. S. Fish and Wildlife Service, Missouri Department of Conservation, University of Missouri, and Wildlife Management Institute cooperating).

REFERENCES

Amir, A. A. 1987. Polyculture economic development with fertilizers in Hungary. Aquaculture Magazine 13(3): 30-33.

APHA et al. (American Public Health Association, American Water Works Association, and Water Pollution Control Federation). 1980. Standard Methods for the Examination of Water and Wastewaters, 15th ed. American Public Health Association, New York, New York.

Bowling, C. W., W. P. Rutledge, and J. G. Geiger. 1984. Evaluation of ryegrass cover crops in rearing ponds for Florida Largemouth bass. Progressive Fish-Culturist 46:55-57.

Bulkley, R. V. 1975. Chemical and physical effects on the centrarchid basses. Pages 286-294 *in* R. H. Stroud and H. Clepper, eds. Black Bass Biology and Management. Sport Fishing Institute, Washington, D.C.

Coutant, C. C. 1975. Responses of bass to natural and artificial temperature regimes. Pages 272-285 *in* R. H. Stroud and H. Clepper, eds. Black Bass Biology and Management. Sport Fishing Institute, Washington, D.C.

Coutant, C. C. 1987. Thermal preference: when does an asset become a liability. Environmental Biology of Fishes 18:161-172.

Doyle, K. M., and C. E. Boyd. 1984. The timing of inorganic fertilization of sunfish ponds. Aquaculture 37:169-177.

Dupree, H. K., and J. V. Huner. 1984. Third Report to the Fish Farmers–The Status of Warmwater Fish Farming and Progress in Fish Farming Research. U. S. Fish and Wildlife Service, Washington, D.C.

Farquhar, B. W. 1984. Evaluation of fertilization used in striped bass, Florida largemouth bass, and smallmouth bass rearing ponds. Proceedings of the Southeastern Association of Fish and Wildlife Agencies 38:346-368.

Geiger, J. G. 1983. Zooplankton production and manipulation in striped bass rearing ponds. Aquaculture 35:331-351.

Geiger, J. G., and N. C. Parker. 1985. Survey of striped bass hatchery management in the southeastern United States. Progressive Fish-Culturist 47:1-13.

Geiger, J. G., C. J. Turner, K. Fitzmayer, and W. C. Nichols. 1985. Feeding habits of larval and fingerling striped bass and zooplankton dynamics in fertilized rearing ponds. Progressive Fish-Culturist 47:213-223.

Gill, J. L., and H. D. Hafs. 1971. Analysis of repeated measurements of animals. Journal of Animal Science 44:331-336.

Hepher, B. 1963. Ten years of research in fishpond fertilization in Israel. II. Fertilizers dose and frequency of fertilization. Bamidgeh 15:78-92.

Hodson, R. G., and K. Strawn. 1968. Food of young-of-the-year largemouth and spotted bass during the filling of Beaver Reservoir, Arkansas. Proceedings of the Southeastern Association of Game and Fish Commissioners 22:510-516.

Ivleva, I. V. 1973. Mass Cultivation of Invertebrates: Biology and Methods. Israel Program for Scientific Translations Ltd., Jerusalem.

Knowlton, M. F. 1984. Flow-through microcuvette for fluorometric determination of chlorophyll. Water Research Bulletin 20:795-799.

McQueen, D. J., and R. S. Lean. 1987. Influence of water temperature and nitrogen to phosphorus ratios on the dominance of blue-green algae in the Lake St George, Ontario. Canadian Journal of Fisheries and Aquatic Sciences 44:598-604.

Noriega-Curtis, P. 1979. Primary productivity and related fish yield in intensely manured fishponds. Aquaculture 17:335-344.

Parmley, D., G. Alvarado, and M. Cortez. 1986. Food habits of small hatchery-reared Florida largemouth bass. Progressive Fish-Culturist 48:264-267.

Pennak, R. W. 1978. Fresh-water Invertebrates of the United States, 2nd ed. John Wiley and Sons, New York, New York.

Piper, R. G., I. B. McElwain, L. E. Orme, J. P. Fowler, and J. R. Leonard. 1982. Fish Hatchery Management. U.S. Fish and Wildlife Service, Washington, D.C.

SAS Institute, Inc. 1985. SAS/STAT Guide for Personal Computers, Version 6 ed. SAS Institute Inc., Cary, North Carolina.

Snedecor, G. W., and W. G. Cochran. 1980. Statistical Methods, 7th ed. The Iowa State University Press, Ames, Iowa.

Sobue, S., N. Castagnolli, and R.A. Pitelli. 1977. Biotic productivity in fish ponds. Revista Brasileira de Biologia 37:761-769.

Welch, E. B. 1980. Ecological Effects of Waste Water. Cambridge University Press, New York, New York.

Wohlfarth, G. W., and G. L. Schroeder. 1979. Use of manure in fish farming: a review. Agricultural Wastes 1:279-299.

Comparisons of Two By-Products and a Prepared Diet as Organic Fertilizers on Growth and Survival of Larval Paddlefish, *Polyodon spathula,* in Earthen Ponds

Steven D. Mims
Julia A. Clark
John C. Williams
David B. Rouse

ABSTRACT. Two agro-industrial by-products, rice bran (RB) and distillers dried solubles (DS), and a prepared diet (PD) were evaluated as organic fertilizers for the production of juvenile paddlefish in nine 0.02-ha earthen ponds over a 40-day culture period. Paddlefish yield from ponds fertilized with RB (209 kg/ha) was significantly greater ($P \leq 0.05$) than that from ponds fertilized with DS (129 kg/ha), but it was not significantly greater than yields from ponds fertilized with PD (258 kg/ha). Fish survival from ponds fertilized with PD (79%) was significantly higher than from ponds fertilized with RB (55%) or DS (50%). There was no significant difference in

Steven D. Mims and Julia A. Clark, Aquaculture Research Center, Community Research Service, Kentucky State University, Frankfort, KY 40601, USA.

John C. Williams, Research Data Analysis, Alabama Agricultural Experiment Station, Auburn University, AL 36849, USA.

David B. Rouse, Department of Fisheries and Allied Aquacultures, Alabama Agricultural Experiment Station, Auburn University, AL 36849, USA.

[Haworth co-indexing entry note]: "Comparisons of Two By-Products and a Prepared Diet as Organic Fertilizers on Growth and Survival of Larval Paddlefish, *Polyodon spathula,* in Earthen Ponds." Mims, Steven D. et al. Co-published simultaneously in the *Journal of Applied Aquaculture,* (The Haworth Press, Inc.) Vol. 2, No. 3/4, 1993, pp. 171-187; and: *Strategies and Tactics for Management of Fertilized Hatchery Ponds* (ed: Richard O. Anderson and Douglas Tave) The Haworth Press, Inc., 1993, pp. 171-187. Multiple copies of this article/chapter may be purchased from The Haworth Document Delivery Center [1-800-3-HAWORTH; 9:00 a.m. - 5:00 p.m. (EST)].

survival between ponds fertilized with RB and DS. Secchi disk visibilities in ponds fertilized with RB were significantly lower than in ponds fertilized with DS and PD. Relatively low Secchi disk visibilities in RB-fertilized ponds were because of a brown stain or coloration which reduced sunlight penetration and growth of filamentous algae, not observed in DS- or PD-fertilized ponds. Larvae congregated in areas where PD was being applied, which suggested direct feeding on PD. Paddlefish did not respond when RB and DS were applied to ponds. Cost per juvenile paddlefish raised in ponds fertilized with RB was $0.004, cheaper than $0.011 for fish raised in ponds fertilized with PD or DS. Rice bran is the recommended agro-industrial by-product to raise juvenile paddlefish greater than 120 mm total length based on improved fish yields, pond water quality, and lower cost per fish. The prepared diet may be used not only as an organic fertilizer, but also as a supplemental feed.

INTRODUCTION

Paddlefish, *Polyodon spathula,* is valued as a commercial fish for both its flesh and its roe and also as a sport fish (Carlson and Bonislawsky 1981). However, paddlefish are listed as a special concern species in 23 states of United States (Williams et al. 1989). Continued loss of its riverine habitat and over-exploitation by commercial fisheries have been the principle threats to endemic paddlefish populations (Williams et al. 1989). Many state and federal fishery agencies are developing mitigation and restoration programs which require the stocking of cultured paddlefish to increase the population in its native range. Paddlefish culture is also being investigated because the fish are considered a desirable food fish (Semmens and Shelton 1986). The flesh is firm and completely boneless, making it popular to consumers (Decker et al. 1991). Proven techniques of raising larvae to juveniles (>120 mm total length) are an integral part of establishing a consistent supply of paddlefish to meet repopulation and culture demands.

Young fish less than 120 mm total length (TL) are particulate feeders and consume large cladocerans, especially *Daphnia* (Michaletz et al. 1982). Large cladocerans must be available during the first two weeks post-stocking for reliable paddlefish production (Mims et al. 1991). Thereafter, older paddlefish greater than 50 mm TL can feed on less preferred food items–small cladocerans and chironomids (Michaletz et al. 1983; Mims et al. 1991).

Various types, quantities, and combinations of organic fertilizers have been used to stimulate zooplankton (primarily cladocerans) production in paddlefish culture ponds and to maintain suitable water quality for growth and development of the fish (Graham et al. 1986). Michaletz et al. (1982) reported that paddlefish larvae stocked at 49,000/ha reached 120 mm TL, with a survival of 16% after 80 days in ponds that were fertilized with a combination of brewer's yeast (453 kg/ha), alfalfa meal (227 kg/ha), dehydrated cow manure (453 kg/ha), and 10 bales of clover hay. Semmens (1982) stocked 49,000 larvae/ha into ponds that were fertilized with alfalfa pellets (600 kg/ha) and meat/bone meal (300 kg/ha) and inoculated with *Daphnia* spp.; after 40 days, he harvested fish that were about 100 mm, and survival averaged 58%. Recently, Mims et al. (1991) reported that ponds stocked at 61,775 larvae/ha, fertilized with rice bran (1,742 kg/ha) and liquid inorganic fertilizer (10-34-0; 69.5 L/ha), and inoculated with *Daphnia pulex* (0.5/L) produced juvenile paddlefish greater than 120 mm TL, with an average survival rate of 77% after 40 days. A high stocking density combined with organic and inorganic fertilization and an inoculation with preferred food organisms increased the number of paddlefish harvested by more than 35% and the yield by more than 128%, compared to results by Michaletz et al. (1982) and Semmens (1982). Further improvements in larval survival and juvenile yields from earthen ponds may be possible by finding organic fertilizers that, when combined with inorganic fertilizer and applied at the proper rates, can increase the production of *Daphnia* spp. and provide more food for the paddlefish.

The objectives of this study were to compare two commercially available agro-industrial by-products and a prepared diet as organic fertilizers to improve paddlefish survival and yield in earthen ponds and to determine the impact of these organic fertilizers on selected water quality variables over a 40-day period.

MATERIALS AND METHODS

Nine 0.02-ha earthen ponds located at Kentucky State University Aquaculture Research Center, Frankfort, Kentucky were used. Three ponds were randomly assigned to each of three treatments.

Ponds received a pre-flooding treatment with liquid Hydrothol[®][1] at a rate of 0.2 mg/L to control filamentous algae. On April 12, 1989, ponds were filled to an average depth of 1.1 m with water taken from a surface water reservoir; the water was filtered through 385-μ saran cloth socks.

Two commercially available agro-industrial by-products–rice bran (RB) and distillers dried solubles (DS)–and a prepared diet (PD)–Purina trout chow starter (#5100)–were evaluated as organic fertilizers. Fertilizer quantities and application schedules for each treatment were based on the nitrogen content of RB (control) as described by Mims et al. (1991) (Table 1). Total amount of nitrogen applied to each pond via organic fertilizers was 43 kg/ha (3,911 μg/L). Organic fertilizers were analyzed as described by Horwitz (1980); carbon to nitrogen (C:N) ratios were determined with an elemental analyzer for macrosample (LECO CHN, model 600, St.Joseph, Michigan) (Table 2). Each pond received an additional 11 kg/ha (933 μg/L) of nitrogen in the form of liquid ammonium polyphosphate (10-34-0) applied at varied amounts over the experimental period (Table 1). The weekly quantities of nitrogen and phosphorus (μg/L) added to the ponds in the form of organic and inorganic fertilizers are reported in Table 3.

Ponds were inoculated with zooplankton, predominately *Daphnia* spp., at a rate of about 125,000 crustaceans per pond on April 13, 14, and 15. Each pond was equipped with one continually operated 5-cm polyvinyl chloride air-lift pump to provide thorough mixing of nutrients and zooplankton during the study (Parker 1979).

Paddlefish larvae, produced from broodfish collected in Kentucky, were held in 50-cm square boxes with window screen (800-μ mesh size) bottoms for 8 days after hatching until mouth parts were well developed, peristalsis had begun, and larvae were actively seeking food (Graham et al. 1986). Larvae averaging 16.8 mm TL (SD = 0.3 mm) and 23.0 mg (SD = 2.1 mg) were stocked on April 27 at 61,775 larvae/ha. Ten paddlefish were collected by seining from each pond weekly. The fish were measured in centimeters for TL and in grams for weight. Fish were harvested after 40 days. An

---

1. Use of trade or manufacturer names does not imply endorsement.

TABLE 1. Organic and inorganic fertilization rates for 0.02-ha ponds that were stocked with 61,775 larval paddlefish/ha. Organic fertilizers were divided into the indicated number of applications. All treatments received the same amount of inorganic fertilizer (10-34-0). Three weekly applications of fertilizers were applied to the filled ponds during a two-week period prior to stocking (week 0).

			Organic fertilizers		
Week	No. of applications	Rice bran (kg/ha/week)	Distillers dried solubles (kg/ha/week)	Prepared diet (kg/ha/week)	Inorganic fertilizer (l/ha/week)
0	6	1,410	600	345	37.0
1	3	310	132	76	4.6
2-5	8	157	67	39	9.3
Total	17	2,348	1,000	577	78.8

TABLE 2. Analysis of organic fertilizers applied to 0.02-ha ponds that were stocked with 61,775 larval paddlefish/ha.

Composition	Rice bran	Distillers dried solubles	Prepared diet
Crude Protein (%)	11.4	27.1	47.0
Fat (%)	14.1	6.2	24.7
Crude Fiber (%)	12.8	3.5	1.1
Moisture (%)	9.5	4.4	7.5
Phosphorus (%)	1.5	1.3	1.6
Potassium (%)	1.5	1.6	0.6
Magnesium (%)	0.8	0.5	0.2
Calcium (%)	1.1	0.1	2.5
C:N ratio	22:1	10:1	6:1

TABLE 3. Weekly nitrogen (N) and phosphorus (P) for organic and inorganic fertilizers, in mg/l applied to 0.02-ha ponds that were stocked with 61,775 larval paddlefish/ha. Ratios were calculated from total N and P from the combination of the organic and inorganic fertilizers.

				Organic fertilizers					
		Rice bran		Distillers dried solubles		Prepared diet		Inorganic fertilizer	
Week	No. of applications	N	P	N	P	N	P	N	P
0	6	2,336	295	2,336	600	2,358	407	467	691
1	3	519	65	519	132	519	89	58	85
2-5	8	1,056	132	1,056	268	1,056	184	488	173
Total	17	3,911	492	3,911	1,000	3,911	680	993	1,468
Ratio		2.5 : 1		2.0 : 1		2.3 : 1			

additional 40 fish were sampled at harvest for final individual TL and weight.

Water Quality Analysis and Management

Dissolved oxygen and water temperature, (polarographic dissolved oxygen meter and thermistor, YSI model 54A) and pH (Omega pHH-43 meter) were measured at 0700 and 1500 daily in each pond. Secchi disk visibility was measured daily at 0700. Water samples were collected weekly from each pond before fertilizers were applied and analyzed for alkalinity (as mg/L of $CaCO_3$), ammonia-nitrogen (NH_3–N), nitrite-nitrogen (NO_2–N), nitrate-nitrogen (NO_3^-–N), and total filterable orthophosphate (PO_4–P) (Hach DREL/5). Chlorophyll *a* was extracted from water samples with acetone and measured spectrophotometrically (APHA et al. 1980). Emergency aeration was provided by a 0.33-hp surface aerator whenever the dissolved oxygen concentration was predicted by graph (Boyd 1979) to decline below 40% of saturation (Andrews et al. 1973). Ponds fertilized with DS or PD were treated with 0.1 mg/L Hydrothol® on week 5 to control filamentous algae. Eleven kg of sodium chloride were added to each pond to prevent possible nitrite-induced anemia (Tucker et al. 1989).

Statistical Analyses

Differences in survival and yields were assessed by analysis of variance (ANOVA) for a completely randomized design (SAS Institute, Inc. 1990). Paddlefish size and water quality data observed weekly in the ponds were assessed by a repeated measures version of ANOVA (split plot design). Mean differences between selected treatment means (*a priori* comparisons) were tested with contrasts. Weekly trends in water quality variables were tested using orthogonal polynomial contrasts. The probability level for tests was 0.05.

RESULTS

Fish Yield and Survival

During week 5, growth of filamentous algae in one DS pond could not be controlled with Hydrothol.® As a result, fish became entangled, and severe mortality occurred although it could not be accurately recorded. Because of low fish yield and survival, data in this pond were excluded from the analysis.

At harvest, mean paddlefish yield in ponds fertilized with RB was significantly greater than in ponds fertilized with DS, although the difference in survival was not significant (Table 4). Mean yield in ponds fertilized with RB were not significantly different than that from ponds fertilized with PD. However, mean survival in ponds fertilized with PD was significantly higher than that from ponds fertilized with RB.

Growth of the fish from the different treatments is shown in Figure 1. Statistical differences in fish TL and weight from ponds fertilized with RB or DS were noted on weeks 5 and 6 after stocking. Fish sampled from ponds fertilized with RB were significantly longer and heavier than fish from ponds fertilized with DS. Fish sampled from ponds fertilized with PD or RB were not significantly

TABLE 4. Mean yield, growth rate, and survival of paddlefish larvae ($\pm$ SE) stocked at 61,775/ha and cultured for 40 days in 0.02-ha ponds fertilized with rice bran, distillers dried solubles, or a prepared diet. Values followed by the same letter in each column are not significantly different at $P > 0.05$.

Treatment	Yield (kg/ha)	Growth rate (mm/d)	Survival (%)
Rice bran	219±22a	2.8±0.2a	55±5b
Distillers dried solubles	129±27b	2.3±0.1b	50±6b
Prepared diet	258±22a	2.6±0.1a	79±5a

different in size at all sampling dates, despite higher survival in ponds fertilized with PD. Total length and weight in each treatment increased linearly over weeks. At harvest, fish sampled in ponds fertilized with RB and PD averaged 130 and 120 mm TL, respectively, whereas DS fish averaged only 105 mm TL.

Water Quality

Mean Secchi disk visibilities for the different treatments are shown in Figure 2. Secchi disk visibilities for the test period in ponds fertilized with RB (61 cm) were significantly lower than those in ponds fertilized with DS (91 cm) or with PD (86 cm). Weekly Secchi disk visibilities in RB ponds were significantly lower than those in PD ponds on weeks 2, 3, and 4 or in DS ponds on weeks 2, 4, and 5. Mean phytoplankton abundance, as measured by chlorophyll-a concentrations, in ponds fertilized with DS (40 μg/L) was significantly higher than that in RB (16 μg/L) and in PD (20 mg/L) ponds. Weekly chlorophyll-a concentrations are illustrated in Figure 3. The low water transparency and low chlorophyll-a in RB ponds were due to a brown coloration or stain imparted by the RB. There were no statistical differences in phosphorous levels among treatments. Because of the water clarity in ponds fertilized with DS or PD, filamentous algae became established.

Mean alkalinity for ponds fertilized with RB (97 mg/L) was significantly higher than those fertilized with DS (87 mg/L) or with PD (84 mg/L). Mean pH for ponds fertilized with PD (8.4) was significantly higher than that in ponds fertilized with RB (7.8) or

FIGURE 1. Mean weekly total lengths (A) and weights (B) of paddlefish in ponds fertilized with rice bran, distillers dried solubles, or prepared diet and the estimated linear increases. Mean total length and weight at stocking were 17 mm and 23 mg, respectively.

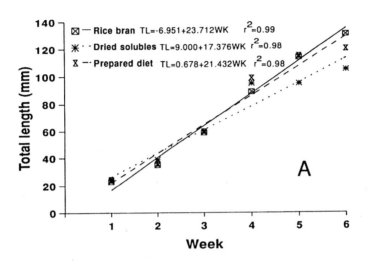

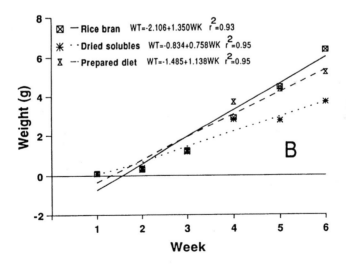

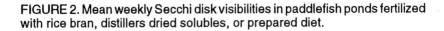

FIGURE 2. Mean weekly Secchi disk visibilities in paddlefish ponds fertilized with rice bran, distillers dried solubles, or prepared diet.

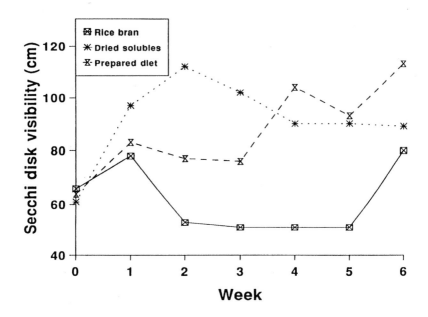

with DS (8.0). Mean weekly pH for the different treatments are shown in Figure 4. Weekly pH in ponds fertilized with RB was significantly lower on weeks 1, 3, and 4 than that in PD ponds, and it was significantly lower on week 4 than that in DS ponds.

Mean early morning dissolved oxygen in ponds fertilized with RB (7.1 mg/L) was significantly lower than that in ponds fertilized with DS (7.7 mg/L); the average dissolved oxygen of RB and DS ponds (7.4 mg/L) was significantly lower than in ponds fertilized with PD (8.6 mg/L). Ponds fertilized with RB received the greatest amount of organic fertilizer (Table 1) and had the lowest weekly dissolved oxygen as shown in Figure 5. Dissolved oxygen in ponds fertilized with DS was significantly higher on weeks 1, 4, and 5 and was higher in ponds fertilized with PD on weeks 1, 3, and 4 than that in RB ponds. Predicted low dissolved oxygen (≤ 40 % of saturation) required emergency aeration on 11, 8, and 4 nights for ponds fertilized with RB, DS, and PD, respectively.

There were no statistical differences in water temperature among

FIGURE 3. Mean weekly chlorophyll-*a* concentrations in paddlefish ponds fertilized with rice bran, distillers dried solubles, or prepared diet.

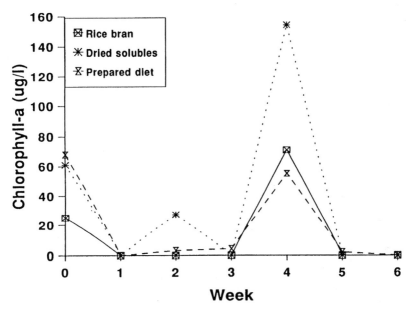

treatments. Water temperature averaged 20.7°C; unseasonally cold mean water temperatures were experienced during week 2 (16.4°C) and week 3 (17.9°C).

Mean ammonia and nitrite values were not significantly different among ponds fertilized with RB, DS, or PD over the culture period or by weeks. However, lower average levels of ammonia (0.12 mg/L) and nitrite (0.004 mg/L) over the culture period were observed for ponds fertilized with RB than those in ponds fertilized with DS or PD. Mean observed levels of ammonia and nitrite over the culture period were 0.31 and 0.16 mg/L for DS-fertilized ponds and 0.24 and 0.19 mg/L for PD-fertilized ponds, respectively.

DISCUSSION

Choice of organic fertilizers in ponds used to raise larval fishes should be based on local availability, cost, personal experience, and zooplankton response (Barkoh and Rabeni 1990). Mims et al.

FIGURE 4. Mean weekly afternoon pH in paddlefish ponds fertilized with rice bran, distillers dried solubles, or prepared diet.

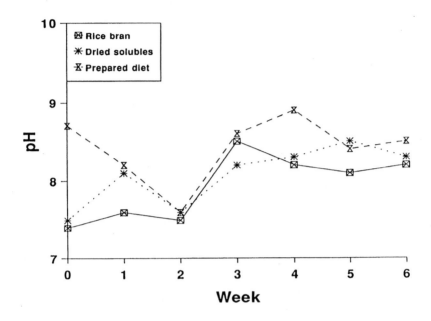

(1991) reported that RB met those criteria and should be used as the recommended organic fertilizer for production of juvenile (≥ 120 mm in TL) paddlefish. Rice bran also serves as a direct food source for cladocera (De Pauw et al. 1981), provides satisfactory and stable water quality, and reduces the incidence of filamentous algae (Mims et al. 1991). In the present study, higher survival and significantly greater yields were obtained in ponds fertilized with RB than those in ponds fertilized with DS. Differences in fish growth rates between the two by-products were found. Fish sampled during weeks 5 and 6 in ponds fertilized with RB had growth rates of 3.6 and 2.3 mm/day, respectively, compared to growth rates of 0 and 1.4 mm/day, respectively, in ponds fertilized with DS. Mims et al. (1991) reported that paddlefish growth rates ≥ 2.5 mm/day indicated availability of preferred food items.

Mean fish yield and survival in ponds fertilized with RB were lower than those previously reported by Mims et al. (1991). Lower

FIGURE 5. Mean weekly morning dissolved oxygen in paddlefish ponds fertilized with rice bran, distillers dry solubles, or prepared diet.

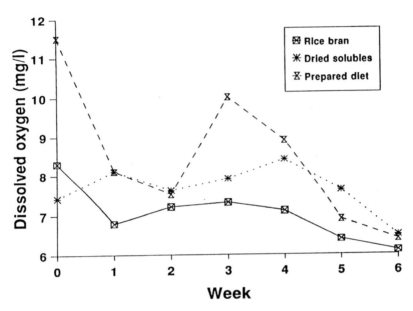

yield and survival in this study probably resulted from: colder water temperatures (20 vs. 24°C) and its effects on decomposition of organic matter by bacteria (Barkoh and Rabeni 1990); zooplankton generation time, longevity and reproduction (Allan 1976); and larval paddlefish growth (Mims and Schmittou 1989). The linear relationships of TL and weight over weeks suggest that maximum standing crops were not reached in any of the treatments. Semmens (1982) stocked 49,000 larvae/ha and fertilized with 600 kg/ha of alfalfa pellets and 300 kg/ha of meat/bone meal and harvested 125 kg of paddlefish/ha with an average TL of 100 mm in 40 days, which were lower and shorter than those obtained in this study.

Fish in ponds fertilized with PD had a greater mean weight per fish than fish in ponds receiving DS. The prepared diet could have served also as a supplemental feed. Paddlefish congregated around the feed and appeared to feed aggressively when PD was broadcast over the water surface. The same feeding response was not observed when RB or DS was applied. A combination of live organisms and prepared diets has resulted in greater larval growth and survival of common carp, *Cyprinus carpio* (Jhingrah and Pullin 1985), bighead carp, *Aristichthys nobilis* (Fermin and Rocometa 1988), striped bass, *Morone saxatilis* (Fitzmayer et al. 1986), and smallmouth bass, *Micropterus dolimieui* (Ehrlich et al. 1989), than have diets consisting either of live organisms or of prepared diets alone. A combination of live organisms and prepared diets has not been tested on larval paddlefish. Fitzmayer et al. (1986) have shown that supplemental feeding increased not only hybrid striped bass yield and survival, but that nutrients released from uneaten feed and fish feces by bacterial decomposition helped maintain zooplankton in the ponds for longer periods. Mims (1992) also reported higher *Daphnia* densities and biomass in PD ponds, compared to those in RB ponds in response to post-stocking fertilization.

The impact of fertilizers on water quality must also be considered when choosing an appropriate fertilizer. In general, water quality was better in ponds fertilized with RB than in ponds fertilized with DS or PD. Ponds fertilized with RB retained a brown coloration or stain during the 40-day period and had lower Secchi disk visibilities, reduced sunlight penetration, and filamentous algae growth compared to ponds fertilized with DS or PD. The reduction in filamentous algae is significant because filamentous algae are known to be detrimental to paddlefish survival (Mims et al. 1991). Alkalinity in ponds fertilized with RB was higher than in ponds fertilized with DS or PD. Higher alkalinity probably occurred because of the high organic load in RB ponds, with its increased bacterial decomposition, and produced greater amounts of carbon dioxide which, in turn, increased the solubility of calcium carbonate (Boyd 1990). All other water quality parameters were within acceptable ranges for paddlefish (Mims et al. 1991).

Rice bran cost $63/MT, DS cost $330/MT, and PD cost $898/MT. Based on quantities applied in this study (Table 1), RB was the cheapest; it cost $148/ha, compared to $331/ha for DS and $518/ha for PD. Despite higher paddlefish survival in ponds fertilized with PD, prices of juvenile fish, based only on the cost of fertilizer, were: $0.004/fish for RB, $0.011/fish for DS, and $0.011 fish for PD. Therefore, juvenile paddlefish production in earthen ponds fertilized with RB was 2.8 times less expensive than production in ponds fertilized with PD or DS.

Experimental results support the findings of Mims et al. (1991) that RB is recommended over DS and PD as an organic fertilizer to stimulate *Daphnia* spp. production in paddlefish ponds, based on fish yields, impact on water quality, and cost. Data and observations indicate that PD may serve not only as a fertilizer to stimulate *Daphnia* spp. production but also as a supplemental feed. Further research should compare intensive larval paddlefish production in ponds, with prepared diets as a direct food source and in ponds with live organisms or a combination of both to increase fish production and survival.

ACKNOWLEDGMENTS

The authors express their appreciation to Richard Knaub and Danny Yancey for their assistance during the study. We also would like to thank Leonard Lovshin for his critical review of the manuscript. Rice bran was donated by Riceland Foods, Inc., Stuttgart, Arkansas, and distiilers dried solubles was donated by Brown-Foreman, Louisville, Kentucky. This research was supported by USDA/CSRS grant to Kentucky State University under agreement KYX-80-85-01A.

REFERENCES

Allan, J. D. 1976. Life history patterns in zooplankton. The American Naturalist 110:165-180.

Andrews, J. W., T. Murai, and G. Gibbons. 1973. The influence of dissolved oxygen on the growth of channel catfish. Transactions of the American Fisheries Society 102:835-838.

APHA et al. (American Public Health Association, American Water Works Association, and Water Pollution Control Federation). 1980. Standard Methods for the Examination of Water and Wastewater, 15th ed. American Public Health Association, Washington, D.C.

Barkoh, A., and C. F. Rabeni. 1990. Biodegradability and nutritional value to zooplankton of selected organic fertilizers. Progressive Fish-Culturist 52:19-25.

Boyd, C. E. 1979. Water Quality in Warmwater Fish Ponds. Alabama Agricultural Experiment Station, Auburn University, Alabama.

Boyd, C. E. 1990. Water Quality in Ponds for Aquaculture. Alabama Agricultural Experiment Station, Auburn University, Alabama.

Carlson, D. M., and P.S. Bonislawsky. 1981. The paddlefish (*Polyodon spathula*) fisheries of the midwestern United States. Fisheries 6(2):17-22, 26-27.

Decker, E. A., A. D. Crum, S. D. Mims, and J. H. Tidwell. 1991. Processing yields and composition of paddlefish (*Polyodon spathula*), a potential aquaculture species. Journal of Agricultural and Food Chemistry 39:686-688.

De Pauw, N., P. Laureys, and J. Morales. 1981. Mass cultivation of *Daphnia magna* (Strauss) on rice bran. Aquaculture 25:141-152.

Ehrlich, K. F., M. C. Cantin, and M. B. Rust. 1989. Growth and survival of larval and post-larval smallmouth bass fed a commercially prepared dry feed and/or *Artemia* nauplii. Journal of the World Aquaculture Society 20:1-6.

Fermin, A. C., and R. D. Rocometa. 1988. Larval rearing of bighead carp, *Aristichthys nobilis* Richardson, using different types of feed and their combinations. Aquaculture and Fisheries Management 19:283-290.

Fitzmayer, K. M., J.I. Broach, and R. D. Estes. 1986. Effects of supplemental feeding habits of striped bass in ponds. Progressive Fish-Culturist 48:18-24.

Graham, L. K., E. J. Hamilton, T. R. Russell, and C. E. Hicks. 1986. The culture of paddlefish–a review of methods. Pages 78-94 *in* J. G. Dillard, L. K. Graham, and T. R. Russell, eds. The Paddlefish: Status, Management and Propagation. Modern Litho-Print Co., Jefferson City, Missouri.

Horwitz, W. 1980. Official Methods of Analysis of The Association of Official Analytical Chemists, 13th ed. Association of Official Analytical Chemists. Washington, D.C.

Jhingrah, V. J., and R. S. V. Pullin. 1985. Hatchery Manual for the Common, Chinese, and Indian Major Carps. International Center for Living Aquatic Resources Management, Manila, Philippines.

Michaletz, P. H., C. F. Rabeni, W. W. Taylor, and T.R.Russell. 1982. Feeding ecology and growth of young-of-the-year paddlefish in hatchery ponds. Transactions of the American Fisheries Society 111:700-709.

Michaletz, P. H., C. F. Rabeni, W. W. Taylor, and T.R.Russell. 1983. Factors affecting *Daphnia* declines in paddlefish rearing ponds. Progressive Fish-Culturist 45:120-132.

Mims, S. D. 1992. Juvenile Paddlefish Production in Earthen Ponds. Doctoral dissertation, Auburn University, Alabama.

Mims, S. D., and H. R. Schmittou. 1989. Influence of *Daphnia* density on survival and growth of paddlefish larvae at two temperatures. Proceedings of the Southeastern Association of Fish and Wildlife Agencies 43:112-118.

Mims, S.D., J. A. Clark, and J. H. Tidwell. 1991. Evaluation of three organic fertilizers for paddlefish production in nursery ponds. Aquaculture 99:69-82.

Parker, N. C. 1979. Striped bass culture in continuously aerated ponds. Proceedings of the Southeastern Association of Fish and Wildlife Agencies 33:353-360.

SAS Institute, Inc. 1990. SAS User's Guide: Statistics, Version 6 ed., SAS Institute, Inc., Cary, North Carolina.

Semmens, K. J., 1982. Production of Fingerling Paddlefish (*Polyodon spathula*) in Earthen Ponds. Master's thesis, Auburn University, Alabama.

Semmens, K. J., and W. L. Shelton. 1986. Opportunities in paddlefish aquaculture. Pages 106-113 *in* J. G. Dillard, L.K. Graham, and T. R. Russell, eds. The Paddlefish: Status, Management and Propagation. Modern Litho-Print Co., Jefferson City, Missouri.

Tucker, C. S., R. Francis-Floyd, and M. H. Beleau. 1989. Nitrite-induced anemia in channel catfish, *Ictalurus punctatus* (Rafinesque). Bulletin of Environmental Contamination and Toxicity 43:295-301.

Williams, J. E., J.E. Johnson, D. A. Hendrickson, S. Contreras-Balderas, J. D. Williams, M. Navarro-Mendoza, D. E. McAllister, and J. E. Deacon. 1989. Fishes of North America endangered, threatened, or of special concern: 1989. Fisheries 14(6):2-20.

Strategies and Tactics for Larval Culture of Commercially Important Carp

Karol K. Opuszynski
Jerome V. Shireman

ABSTRACT. This paper reviews state-of-the-art techniques for culture of larval common carp, *Cyprinus carpio,* silver carp, *Hypophthalmichthys molitrix,* bighead carp, *H. nobilis,* and grass carp, *Ctenopharyngodon idella.* Water temperature, food, and predation are important factors influencing larval survival and growth. Lower and upper lethal temperatures range from 3 to 44°C. Optimum growth temperatures range from 38 to 40°C. Lethal and optimum temperatures vary with acclimation temperature, fish age, and development stage of fish. Water temperatures are close to optimum for larval culture in tropical regions but are often too low in temperate climates. Intensive culture in temperature-controlled systems is important in temperate climates. The first food eaten by larvae in ponds consists mainly of protozoa, rotifers, and copepod nauplii. As the larvae grow, they quickly shift to larger food items, including cladocera and insect larvae. Management practices to enhance natural food development in earthen ponds include bottom drying, soil preparation, liming, fertilization, and agricultural crop cultivation. The development of food organisms in freshly filled ponds follows a pattern of succession. For best results, a pond should be stocked at the stage of succession when the size relationship be-

Karol K. Opuszynski and Jerome V. Shireman, Department of Fisheries and Aquatic Sciences, Institute of Food and Agricultural Sciences, University of Florida, Gainesville, FL 32606, USA.

[Haworth co-indexing entry note]: "Strategies and Tactics for Larval Culture of Commercially Important Carp." Opuszynski, Karol K., and Jerome V. Shireman. Co-published simultaneously in the *Journal of Applied Aquaculture* (The Haworth Press, Inc.) Vol. 2, No. 3/4, 1993, pp. 189-219; and: *Strategies and Tactics for Management of Fertilized Hatchery Ponds* (ed: Richard O. Anderson and Douglas Tave) The Haworth Press, Inc., 1993, pp. 189-219. Multiple copies of this article/chapter may be purchased from The Haworth Document Delivery Center [1-800-3-HAWORTH; 9:00 a.m.- 5:00 p.m. (EST)].

tween fish larvae (predators) and zooplankton (prey) is proper. A common practice is to stock larvae 3-7 days after filling. If ponds are filled too long before larvae are stocked, food relationships between fish and invertebrates can be reversed. Predator control includes biological, chemical, physical, and mechanical methods. Although great progress has been made in the development of dry starter diets, prepared feeds are not yet available for successful large-scale production. This problem is usually overcome by starting larvae with live food or with a mixture of live food and dry feed and by shifting larvae to dry diets as they grow. Live food either is collected from zooplankton ponds or is produced in intensive culture conditions. Systems for larval culture can range from ponds to intensive culture with water recirculation systems. Choice of the best system depends on the local climate, technical, and socio-economic conditions.

INTRODUCTION

Propagation of commercially important fish species has been successful due to recent progress in the development of production techniques. However, heavy larval mortality and slow growth rates during the initial culture period are still problems. This paper describes management techniques for grass carp, *Ctenopharyngodon idella,* silver carp, *Hypophthalmichthys molitrix,* and bighead carp, *H. nobilis.* These species are referred to as Chinese herbivorous carps when all three species are being discussed.

Chinese herbivorous carps have been recently introduced as food fish or for aquatic weed control in Europe, North and South America, and Africa. Chinese herbivorous carps are of great importance in the world fisheries. The silver carp catch was 1,359,724 metric tons, which ranked this fish first in the total world catch in inland waters in 1989. Silver carp comprised 12%, grass carp 8%, and bighead carp 6% of the total catch in inland waters in 1989 (FAO 1991). The techniques, guidelines, and practices for culture of these fish have general applicability to other warmwater fish species. In instances where data on specific techniques could not be found for these species, examples are used which pertain to common carp, *Cyprinus carpio.* Common carp is an important species in Asiatic and European fisheries, and a great amount of data and experience are available.

FACTORS INFLUENCING LARVAL SURVIVAL
AND GROWTH

A knowledge of larval environmental requirements is a basic pre-condition for successful larval culture techniques: important abiotic factors include temperature, oxygen, pH, ammonia, and nitrite; biotic factors include food and predators. Because larval development and growth are rapid during the first weeks after hatching, environmental requirements change during this period. Therefore, the needs of larvae as a function of development and growth must be taken into account during all culture periods.

Temperature

Lower and upper lethal temperatures and optimum temperatures are important concerns for larval culture. Despite detailed studies on the lower lethal temperatures (LLT) of larval Chinese herbivorous carps (Lirski and Opuszynski 1988a), it is difficult to determine precisely the low temperature that kills fish, because LLT depends on the age and size of the fish, previous thermal adaptation, and experimental protocol. Additionally, fish responses to low temperatures are complex, and mortality is often delayed.

When temperatures were decreased at a rate of 0.1°C/minute, D30 grass carp (D1 is day of hatch) tolerated lower temperatures than did D4 fish (Lirski and Opuszynski 1988a). The increase in tolerance, however, was relatively small (0.01-0.07°C/mg body weight). An increase in the acclimation temperature by 3°C caused a 1°C increase in LLT at both ages. The ultimate lower lethal temperature (ULLT), i.e., the highest lower lethal temperature possible, was 17°C for Chinese herbivorous carps. LLT for fish acclimated to 25°C, which is close to the optimum temperature for reproduction for these species, ranged between 9 and 6°C for D4 and D30 fish, respectively (Figure 1).

Grass carp upper lethal temperatures (ULT) increased as the acclimation temperature increased up to 34°C. An increase in the acclimation temperature of 3°C resulted in an increase of the lethal temperature of 1°C (Lirski and Opuszynski 1988b). The ultimate upper lethal temperature (UULT) is of practical importance for

FIGURE 1. Lower lethal temperature (LLT) and upper lethal temperature (ULT) for D4 (dashed line) and D30 (solid line) grass carp as a function of acclimation temperature. Ultimate upper lethal temperature (UULT) is determined as the point where the ULT line crosses the diagonal 45° line (d). Data are from Lirski and Opuszynski (1988a, 1988b).

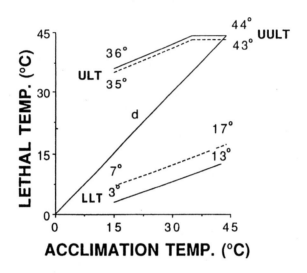

larval culture. UULT is the lethal temperature which is equal to the acclimation temperature (Figure 1). Knowledge of UULT is important for larval culture, because of the general relationship that exists between optimum growth temperature (OGT) and UULT (Jobling 1981): UULT = 0.76 OGT + 13.81.

Once UULT, which is relatively easy to determine is calculated, OGT can be calculated. UULT's for common carp and Chinese herbivorous carps are similar and range from 43 to 44°C (Lirski and Opuszynski 1988b). Consequently, OGT's are 38-40°C. ULT changes little from D4 to 50 (0.01-0.03°C per day) and can be ignored for practical purposes (Opuszynski et al. 1989a).

OGT calculated above is higher than the temperatures that have been used to culture common carp larvae. Literature data differ considerably in this respect. Baranova (1974) gave 23°C as the lower end of the optimum temperature range for growth. Suzuki et al. (1977) found that growth rate increased as temperature increased from 15 to 25°C; an increase from 25 to 30°C did not result in any difference in growth rate. Jowko et al. (1981) reported that temperatures between 28 and 32°C were too high for carp larvae. Some authors stated higher optimum temperatures: 28-32°C (Ostroumova et al. 1980); 30°C (Penaz et al. 1983); 30-34°C (Kapitonova 1977); 36°C (Tatarko 1970). Data are limited for Chinese herbivorous carps. Schlumpberger (1980) reported the thermal optimum for growth of silver carp to be between 25 and 30°C; Vovk (1976) found that it was 32°C. Opuszynski et al. (1989a) found that the weight of silver carp cultured at 35°C tripled in comparison to silver carp that were cultured at 25°C.

In addition to increasing growth rate, raising carp larvae at higher water temperatures protects them against *Saprolegnia* and *Ichthyophthirius*, which are among the most dangerous and frequently occurring diseases (Littak et al. 1980; Wozniewski et al. 1980). These organisms do not develop above 30°C (Olah and Farkas 1978; Prost 1980). Raising carp larvae at higher water temperatures than those mentioned above should be tested. However, since the estimated optimum growth temperatures are close to upper lethal temperatures, accurate temperature control and efficient water aeration would be needed. Further experiments at close-to-lethal temperatures are needed in order to gain more experience and to make this culture method fully applicable.

Oxygen

Lethal oxygen levels (LOL) for larval Chinese herbivorous carps and common carp are similar (Wozniewski and Opuszynski 1988). The lethal oxygen value for D3 silver carp larvae was about 1.8 mg/L, when determined in a 30-minute-test at 25°C. This value decreased sharply with fish age and leveled off below 0.8 mg/L for fish that were older than D20 (Figure 2). These results are similar to those reported for common carp by other authors. Kuznetshova (1958) determined that newly-hatched larvae died when the oxygen concentration was 1.33 mg/L, while D21 fish died when oxygen concentration was 0.9 mg/L. Ashkierov (1975) found an LOL of 1.58, 1.14, and 0.80 mg/L for D1-D6, D8-D14, and D21-D38 fish, respectively. Temperature, at least within the range of 10 to 30°C, did not seem to have a significant effect on LOL (Skhorbatov 1963; Klasthorin and Titov 1975).

No differences in common carp larval survival were found when

FIGURE 2. Relationship between age of silver carp and lethal oxygen concentration (LC). LC is an oxygen level where 50% fish mortality occurred during a 30-minute test at 25°C. The vertical bars show the standard errors (after Wozniewski and Opuszynski 1988).

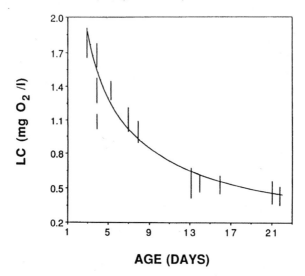

AGE (DAYS)

Reprinted with permission from: Progowa zawartosc tlenu w wodzie dla mlodocianych stadiow ryb karpiowatycyh. Roczniki Nauk Rolniczych 1988 S. H. T. 101 Z. 4

fish were cultured at 33°C and oxygen saturation was between 23 and 168% (Wozniewski and Myszkowski 1987); survival ranged from 89 to 93%. However, significant differences in growth rate were found at different saturation levels. The average weight of fish after a 14-day culture period was 44 mg at 23% saturation, 64 mg at 43-61% saturation, and 80-87 mg at 120-168% saturation levels.

Nitrogen

Nitrogen supersaturation is a real danger in larval carp culture. It can easily occur when water is heated, especially if heating is combined with changing water pressure due to pumping. Larvae of Chinese herbivorous carps are extremely sensitive to nitrogen gas supersaturation during the first few days of exogenous feeding;

total mortality can occur during a few hours of exposure (Wozniewski et al. 1980).

Food

During the transition period, when the larvae weigh between 0.9 and 1.7 mg and transfer from endogenous to exogenous food, they are susceptible to starvation and must find suitable food. This is a critical period in a fish's life. Fifty percent of larvae during the transition stage may die during a 10-day starvation period at 25°C; only 50 to 75% of those that survive will be able to feed when the proper size of zooplankton is abundant (Horoszewicz 1974; Wolnicki and Opuszynski 1988). Fish weighing 5-6 mg, however, endure the same starvation period without mortality (Table 1).

The first food of Chinese herbivorous carps are zooplankton. The initial food of grass carp larvae in a pond was rotifers (Figure 3). These larvae soon switched to copepod nauplii and then copepodits. After 21 days, larvae fed mainly on cladocera and chironomid larvae (Okoniewska and Opuszynski 1988).

Successive changes to larger food items is characteristic for growing larvae (Figure 4). From an energy standpoint, this is of vital importance for the growing larvae in order to optimize energy gain per unit of prey handling time. This follows the "optimal

TABLE 1. Influence of a 10-day-starvation period at 25°C on common carp, silver carp, grass carp, and bighead carp larvae of different weights (data from Horoszewicz [1974] and Wolnicki and Opuszynski [1988]).

	Larvae after yolk resorbtion (0.9-1.7 mg)	5-6 mg larvae
Mortality (%)	50	0
Percent of fish able to commence feeding	50-75	100
Decrease in swimming activity (%)	50-85	0
Loss of weight (%)	35-39	12

FIGURE 3. Percentage by weight of particular food items in the food contents of grass carp. The pond was stocked with D4 larvae on day 0 (after Okoniewska and Opuszynski 1988).

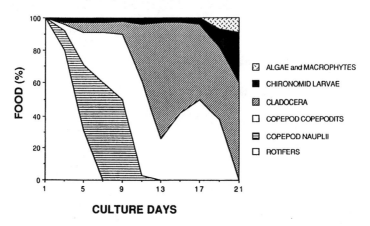

CULTURE DAYS

Reprinted with permission from: Podchow wylegu amura bialego w stawach zasilanych woda podgrzana. Roczniki Nauk Rolniczych 1988 S. H. T. 101 Z. 4

foraging" model postulated by Krebs (1978). The best strategy to achieve this goal is to capture and eat the maximum size prey possible.

During early developmental stages, the maximum size of food organisms is related to mouth size (Shirota 1970; Ito and Suzuki 1977; Dabrowski and Bardega 1984). Carp larvae can be rated in the following order, according to decreasing mouth size: common carp, grass carp, bighead carp, and silver carp (Lirski et al. 1977). For this reason, rotifers and copepod nauplii are not the prey of choice for early stage common carp larvae (Trzoch-Szalkiewicz 1971, 1972; Okoniewska and Wolnicki 1988). These organisms, however, constitute the most important starter food for Chinese herbivorous carps (Lupatsheva 1967, 1970; Panov et al. 1969; Van der Wind 1979). Silver carp larvae feed also on protozoa (Kornenko 1971) and depend on the smallest food items longer than the other species (Ciborowska 1972).

The ratio of larval mouth size to prey size is not the only factor that governs prey size selection. Prey density is also important.

FIGURE 4. Average size of zooplankton organisms eaten and average weight of grass carp as a function of age. The pond was stocked with D4 larvae on day 0 (after Okoniewska and Opuszynski 1988).

CULTURE DAYS

Larvae preferentially capture larger prey at higher prey densities. At low densities, capture of any prey that is encountered is, energetically, the best strategy. Khadka and Rao (1986) found that there was an increase in prey size selection by common carp larvae when zooplankton density increased.

The lack of proper dry starter foods for larval fish is a problem in large-scale culture of cyprinids. Contrary to salmonids, which appear to have a functional stomach before changing from endogenous to external food, cyprinids remain stomachless throughout life and do not have a structurally and functionally differentiated alimentary tract at the time of first feeding. Stomachless fish pose the greatest problems with digestion and assimilation of starter feeds (Dabrowski and Culver 1991).

Great progress in cyprinid larval diet formulation has been made, based on single cell protein (SCP) and freeze-dried animal tissues

(Charlon and Bergot 1984; Charlon et al. 1986). Common carp larvae fed exclusively with artificial food based on yeast cultivated on petroleum by-products and on beef liver grew to an average weight of over 100 mg in 21 days; survival was 87% (Charlon et al. 1986). However, Dabrowski and Poczyczynski (1988) stated that the growth rate of common carp larvae fed on live food was still much better. For example, Vanhaecke and Sorgeloos (1983) using *Artemia* grew common carp larvae to 188 mg in 14 days. Kamler et al. (1990) fed common carp larvae with different artificial diets and found that growth rate, survival, and developmental rate were affected by type of food. They suggested that developmental rate should be considered along with growth rate and survival in the assessment of a dry food.

Recent results suggest that a breakthrough might be imminent in diet formulations for cyprinid larvae. Alami-Durante et al. (1991) found that after 21 days, survival was 95% and mean body weight was 189 mg for common carp larvae fed a yeast-liver diet supplemented with a mineral and vitamin premix. After 10 additional days when they were fed a commercial trout feed, survival was 95% and mean body weight was 755 mg.

Ostroumova et al. (1980) reported excellent results with the Russian starter food Ehkvizo.[1] Common carp gained an average of 1 g after 21 days. Ostroumova and Turestskij (1981) described the use of Ehkvizo for large-scale culture of larval common carp. Recent information from the Inland Fisheries Institute, Poland (J. Wolnicki and W. Gorny, pers. comm.) was that tench, *Tinca tinca,* larvae grew faster when fed Ehkvizo than other commercial starter diets for cyprinids, including Ewos larvstart C 20 (Sweden) and Tetrawerke AZ 30 (Germany).

To date, the difficulty of getting cyprinid larvae to accept dry starter diets are usually overcome by using live food or a mixture of live and dry food initially and by then weaning the fish onto dry food after 1-2 weeks. Bryant and Matty (1981) and Dabrowski (1984) determined that the lowest weight at which larvae could be transferred from live food to an artificial diet was 5-6 mg. Opuszynski et al. (1989b) tested different dry feeds and found that commer-

---

1. Use of trade or manufacturer names does not imply endorsement.

cial trout starter gave results comparable to Ewos C 10 when both feeds were administered after an initial 10-day period when larvae were fed zooplankton.

Predation and Food Competition

Although predation may cause high mortality of carp larvae, few studies have addressed this problem. The most common invertebrate predators are copepods (Sukhanova 1968), and larval and adult insects (Antalfi and Tolg 1975). Among vertebrate predators or competitors for food are tadpoles, adult frogs, and other species of fish (Savin et al. 1973). Opuszynski et al. (1989b) found almost 100% mortality of larval silver carp during the first 6 days of culture in tanks when adult cyclopoid copepods were accidentally introduced with zooplankton collected from ponds. Survival of silver carp larvae was as low as 2% in a pond heavily infested with threespine stickleback, *Gasterosteus aculeatus*; survival ranged from 29 to 36% in ponds where only a few threespine stickleback were found (Opuszynski 1979). Wolny (1970b) found a negative correlation between water transparency and survival of common carp fry in ponds. He felt that low water transparency made it difficult for sight feeders such as birds to prey upon the fry.

INTERACTIONS OF TEMPERATURE, FOOD, AND PREDATION

Several years of experiments with larval carp culture in ponds at the Inland Fisheries Institute, Zabieniec, Poland, made it possible to analyze the influence of temperature, food, and predation on the survival of larvae. All experimental ponds were similar in respect to size (0.2 ha), mean depth (1 m), and water source (a river). All ponds were stocked each year with silver carp larvae from one spawn.

This analysis is based on the following assumptions: (1) if larval survival in a given season is low in all ponds, the influences of temperature, food, and predation on fish survival cannot be separated; (2) if survival is low in some ponds and high in others, temperature can be eliminated; (3) if survival varies among the ponds and at the same time shows a positive correlation with fish growth at the beginning of the season, then a shortage of food may be a contributing factor. It is likely that poor feeding conditions effect fish mortality mainly through predation; slow growth makes larvae vulnerable to predators for a longer period. For this reason, it is difficult to separate these two factors in a pond experiment. In order to separate the influence of feeding conditions and predation, silver carp larvae were also raised in cages. Temperature was increased in some cages with an electric heater and thermostat to determine the influence of temperature (Lirski et al. 1979).

In central Europe, temperature can cause total mortality of silver carp larvae in exceptionally cold seasons; in 1976, 100% mortality occurred in cages with uncontrolled temperature, while over 36% of the larvae survived in the temperature-controlled cages situated in the same pond (Table 2). This substantial difference in survival rate was caused by a 4°C difference in temperature. However, the use of heated cages is impractical when electricity is expensive (Lirski et al. 1977). When predation was eliminated, survival would approach 100% if temperature and food conditions were appropriate (Table 2, 1973 and 1975).

OPTIMIZATION OF ENVIRONMENTAL CONDITIONS FOR LARVAL CULTURE

Temperature

Although natural water temperatures are nearly optimum for the culture of larvae of Chinese herbivorous carps in sub- and tropical climates, water temperatures are a major concern in temperate climates. Pond culture is widely used in temperate climates, even though high mortality might ensue. Moderate or satisfactory survival rates can be obtained during seasons when temperature is average and above average, but total loss of larvae can occur during cold seasons (Table 2). Measures to counteract this problem are limited. Ponds should be stocked late enough to avoid low and highly variable spring temperatures, but early enough and at a density to allow

TABLE 2. Causes of silver carp fry mortality in ponds and cages at the Inland Fisheries Institute, Zabieniec, Poland (+ denotes the cause). Data are from Opuszynski (1979).

Year	Causes of mortality			No. of	Mean survival
	Temperature	Food	Predation	trials	and range (%)
Ponds 1965	+	+	+	3	6(0-10)
1968		+	+	2	69(37-100)
1971	+	+	+	5	5(1-17)
1972		+	+	11	40(2-100)
Cages 1973				2	100
1974	+	+		8	1(0-2)
1975				5	100
1976[a]	+			2	0
1976[b]				3	36(16-64)[c]

[a] Cages with uncontrolled temperature (the lowest recorded temperature = 14°C).

[b] Cages with controlled temperature (the lowest recorded temperature = 18°C).

[c] Some fry escaped from the cage.

growth of the fish to the size that will enable them to survive over the long winter period.

The growth period for larvae of Chinese herbivorous carps lasts only three months (June to August) in temperate climates. Grass carp and silver carp fingerlings must grow to over 4 g and 10 g, respectively, and accumulate enough fat to successfully survive the winter (Opuszynski and Okoniewska 1969). For this reason, the proper stocking density is about 50,000-100,000 larvae/ha. Yield under such conditions usually does not exceed 300 kg/ha (Opuszynski 1969, 1979; Wolny 1970a).

FIGURE 5. Mean water and air temperatures during larval grass carp culture, Goslawice, Poland. Ponds were stocked on 21 June 1980; 1 = heated-effluent inflow water; 2 = flow-through ponds; 3 = static-water ponds; 4 = air temperature. Data are from Okoniewska et al. (1988).

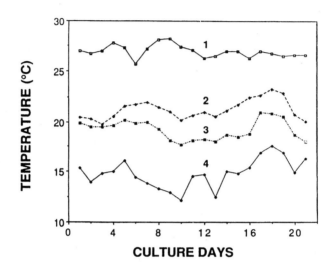

Use of heated effluents from power stations for cyprinid larval culture in earthen ponds is a widespread procedure in Hungary, Germany, Poland, and the Soviet Union where there are numerous power stations with open-water cooling systems. Effluent temperature is usually high enough to increase the temperature in flow-through ponds when the exchange rate is about 2 days (Figure 5). This increase in water temperature can result in a substantial improvement in production. Grass carp larval survival and weight in the static and heated-effluent ponds were 2% and 206 mg, and 63% and 363 mg, respectively, after 21 days of culture (Okoniewska et al. 1988).

Heating of water for larval raising is a cost-effective procedure in cold climates, because densities of 50 to 300+ larvae/L may be held for 1 to 3 weeks. Tanks or aquaria are used, and water quality is maintained either by a flow-through or by a recirculating system. Rottmann et al. (1991) reported that larval culture can be done without water exchange when a simple and inexpensive airlift sponge is employed as a biological filter. This method, if proved feasible in large-scale culture, is advantageous, as the cost of heating water is low and food is not washed from the tanks.

The integration of indoor tanks and outdoor cage culture is an attractive possibility in temperate climates because the growing season can be extended from April until October. This combined and integrated culture system is used in the Netherlands (Huisman 1981). Larvae are raised in temperature-controlled tanks in a greenhouse and then are transferred to cages situated in the cooling water discharge canal of an electric power station.

Food

Three methods are used to feed larvae with natural food: management of natural food in stocked ponds; zooplankton production in ponds without fish; culture of live food.

Stocked ponds

The culture of cyprinid larvae in earthen ponds is by far the oldest, most common, and widely practiced method. It was first employed in China more than 2,400 years ago. Management practices in fry ponds are similar under temperate (Opuszynski 1987) and tropical (Lannan et al. 1986) conditions. Pond preparation includes drying and preparation of the bottom, liming, and fertilization.

Pond drying is of great importance. Fry ponds are kept dry in Europe from late summer until the following spring. The pond bottom is plowed and limed when the ponds are dried. Either 1,000 to 2,500 kg/ha of quicklime (CaO) or 3,000+ kg/ha of slaked lime [Ca(OH)$_2$] is used. In order to increase productivity, agricultural crops such as rye, oats, barley, or clover may be cultivated. The young vegetation may be harvested and used to feed livestock; more often, the vegetation is plowed under before flooding. This is a form of organic fertilization that improves soil structure of the pond bottom.

Fertilization usually increases fish production only in ponds where small algae dominate (Januszko et al. 1977). The best way to enhance primary production and fish yield has been a subject of

FIGURE 6. Common carp yield (kg/ha) after 31 days of culture in unfertilized and chemically fertilized ponds. Each column is the average of two ponds. Data are from Wolny (1970b).

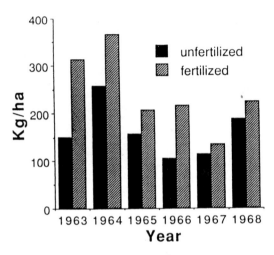

study, discussion, and controversy. In most cases, chemical fertilization increased fish production (Figure 6); however, examples can also be shown where a decline in fish yield occurred (Fijan 1967).

It is generally accepted in Europe that both nitrogen and phosphorus are needed. Recommended N:P ratios range from 4-8:1 (Vinberg and Lakhnovitch 1965) or 2-11:1 (Aleksandrijskaia 1978). Many forms of ammonium fertilizers are used. Piotrowska-Opuszynska (1984) reported the superiority of urea over other chemical fertilizers in common carp ponds. Release of carbon dioxide during hydrolysis helps lower pH; gradual decomposition of urea prevents high ammonia concentrations immediately after fertilization. Superphosphate is the primary source of phosphorus in Europe.

Wolny (1970b) recommended the following recipe for the fertilization of ponds used to culture larval common carp in Central Europe. Total nitrogen fertilizer equal to 210 kg N/ha should be applied in eight applications. Two applications of 4,500 μg N/L each should be applied 6 and 3 days before the pond is stocked. The remaining 6 applications of 2,000 μg/L each are applied at 5-day intervals beginning the day after larvae are stocked. Total superphosphate equal to 23.5 kg P/ha should be applied in four applications. An application of 900 μg P/L is applied 6 days before ponds are stocked. The remaining three applications of 500 μg P/L each are applied 2, 12, and 22 days after stocking. This recipe was created for a culture period of 31 days.

When Wolny's (1970b) recipe is calculated on a weekly basis, the post-stocking fertilization rates are 3,000 μg N/L/week and 375 μg P/L/week. These fertilization rates are high compared to those used in larval fish ponds in the United States. Culver et al. (1993) recommended an N:P fertilization ratio 20:1 and fertilization rates of 600 μg N and 30 μg P/L/week, respectively. Anderson (1993) felt that 600 μg N/L/wk might be too high for striped bass larvae which are highly sensitive to un-ionized ammonia.

Various locally available plant wastes, animal manures, and night soil (in Asia) are used as organic fertilizers. In the U.S., hay, soybean meal, cottonseed meal, and alfalfa meal are also used. The superiority of either organic or chemical fertilization is still a controversial issue. Organic fertilization decreases pH and provides a direct source of food for zooplankton. Increased pH is a common problem in larval ponds which are heavily fertilized with chemical fertilizers. Organic fertilizers, however, can result in oxygen depletion when excessive quantities of organic matter are used.

Lewkowicz and Lewkowicz (1976) compared the effectiveness of chemical and organic fertilizers in ponds with larval common carp. Similar increases in fish production (58-83 kg/ha; about a 50% increase) occurred when chemical fertilizers or cow manure were used. Schroeder et al. (1990) also produced similar fish yields in chemically fertilized or organically manured ponds. Organic matter of the manure contributed only marginally to fish growth. In contrast, Knud-Hansen et al. (1991) showed that fish obtained organic carbon from both primary production and manure-derived detritus when the ponds were fertilized with chicken manure.

The time when larvae should be stocked after a pond is filled is an important management decision. After a pond is filled, different aquatic organisms develop and dominate. This phenomenon is called "biological succession." Long-term studies showed that the biological succession in all types of ponds develops in a similar manner and with the same pattern (Grygierek and Wasilewska 1979). Even though in adjacent ponds different species may dominate and community dynamics as well as fish production may differ, the succession is similar.

The pattern of the succession is shown in Figure 7. With filling, bacteria, heterotrophic protozoa, and heterotrophic algae (such as Euglenoidea) dominate. Autotrophic phytoplankton dominate next. If larval culture is to be successful, ponds should be stocked at the stage when mass-development of rotifers occurs. Stocking the ponds either too early or too late may result in larvae starvation (Opuszynski 1979). If stocking is too early, the zooplankton community is not well developed, and if stocking is too late, the zooplankton may be too large for the larval fish. Additionally, populations of predators may have developed. Rate of succession depends on temperature. A common practice is to stock D4 larvae in ponds that have been filled for 3-7 days.

Proper density of fish is also an important management decision. Stocking density must be adjusted to food resources. These resources greatly surpass the fish demands at the time of stocking but are soon in short supply as fish biomass rapidly increases. For this reason, the culture period in larval ponds is short, and the stocking rate is relatively low (Table 3).

While polyculture of Chinese herbivorous carps is a common

FIGURE 7. General pattern of the succession of organisms in ponds at 15 to 20°C (after Grygierek and Wasilewska [1979]): 1 = bacteria; 2 = protozoa; 3 = phytoplankton; 4 = rotifer; 5 = plankton crustacean.

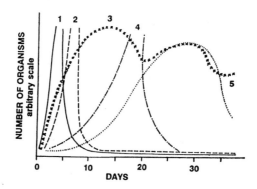

TABLE 3. Stocking rates and culture systems used to raise carp larvae in ponds. Monoculture means that either grass carp, silver carp, or bighead carp were cultured.

Country	Stocking density (x 10^3/ha)	Type of culture	Final fish size (cm)	Culture period (days)	Reference
Poland	150 150	common carp + silver carp	5-7 4-6	30	Opuszynski (1979, 1989); Wolny (1970b)
	150 75	common carp + grass carp, or + bighead carp	5-7 5-7 4-7	30	
China	1,500- 2,500	monoculture	2-3	15-30	FAO (1983)
China	1,000- 1,500	monoculture	3-8	15-30	Marcel (1990)
USA	250- 1,250	monoculture	8-10	21-28[a]	Rottmann and Shireman (1990)
Taiwan	2,000- 3,000	monoculture	2-3	21	Chen (1990)

[a]Data not presented in the paper.

practice in grow-out ponds, monoculture practices prevail in larval ponds. The feeding habits of the early larvae of these species are similar, and sorting and grading are burdensome and may cause increased mortality. Two-species larval polyculture is sometimes used with good results: common carp and silver carp or common carp and grass carp can be raised together (Wolny 1970b; Opuszynski 1979).

While larvae of Chinese herbivorous carps are not fed in ponds in Europe, this is a common practice in Asia and the U.S. In order to make supplemental feeding successful, high stocking densities and high water temperatures ($\geq 22°C$) are needed. Fermented and pre-digested soybean milk is applied several times per day to nursery ponds in Asia. This procedure begins after ponds are stocked with D4 larvae. Soybean milk, however, encourages zooplankton development in ponds rather than nourish the larvae. Dry feed, e.g., minnow meal or catfish starter, is used to feed larvae of Chinese herbivorous carps in the U.S. Feeding begins some time after stocking when larvae are larger and quickly learn to ingest the dry diet.

Zooplankton ponds

Zooplankton can be produced in ponds, harvested, and subsequently delivered to fish-culture facilities as an alternative to larval culture in ponds. Heavy fertilization is used to promote zooplankton development. Such heavy fertilization cannot be used in ponds stocked with larval fish because good water quality standards must be preserved. Zooplankters are resistant to low oxygen concentrations and high ammonia levels. Lincoln et al. (1983) reported that the rotifer *Brachionus rubens* became abundant in ponds when the un-ionized ammonia concentration declined below 20 mg/L. Sewage purification ponds can be used to produce live fish food because mass development of zooplankton often occurs there (Schluter and Groeneweg 1981). In most cases, however, special zooplankton culture ponds that are situated close to the larval culture facilities are used.

In Europe, the same management practices that are used for ponds stocked with larval fish are used in zooplankton culture ponds. The major difference is in fertilization regimens. These procedures may differ, depending on local recipes and available fertilizers, but high doses of organic fertilizers and chemical fertilizers are usually used. Before flooding the ponds, a composted material consisting of a mixture of plant material, animal manures, and 1% quick lime is mixed with chicken manure and is applied at a rate of 7,500 kg/ha. To prevent it from floating after the pond is filled, this mixture is disked into the pond bottom. Liquid cow or pig manure can also be applied at 15,000 kg/ha while the pond is being filled. In addition, chemical fertilization is usually used. Nitrogen and phosphorus fertilizers are applied at concentrations of 5,000 μg N/L and

1,500 μg P/L. Three or four doses are applied at 3-day intervals; the first dose is applied when the pond is being filled.

Rotifers must occur in peak numbers in zooplankton culture ponds when larvae commence exogenous feeding. It is not easy to synchronize these events. The use of several zooplankton culture ponds that are filled at 3-day intervals is the best way to ensure that adequate food exists when it is needed.

Chemicals must be used to kill existing organisms in ponds that cannot be completely drained. In Europe, these chemicals include copper sulphate, liquid ammonia, and organophosphate insecticides (Grygierek and Wasilewska 1979). The last group is of special interest, because of their selective action and short degradation time (Tamas 1979). The organophosphate insecticides Dylox, Flibol, Foschlor, and Neguvon kill crustaceans and encourage rotifer development. The use of these chemicals in concentrations that are needed to kill crustaceans (1,000 μg of active ingredient/L) are safe to fish larvae and can be used in the ponds that are stocked.

Opuszynski and Shireman (1988) found up to 15,000 rotifers/L in Dylox-treated ponds compared to less than 300 rotifers/L in untreated ponds. The first rotifer peak occurred 9 days after the Dylox treatment at 24°C. Opuszynski (1979) found that silver carp larvae grew to an average weight of 285 mg during 30 days in Foschlor-treated ponds but grew to only 57 mg in untreated ponds. The repeated use of organophosphate pesticides prevents the natural succession and aging of zooplankton communities; therefore, the same ponds can be used for rotifer production several times during a growing season.

Collecting zooplankton from the zooplankton ponds is labor intensive in large-scale larval culture. Different zooplankton collectors can be used, but all have the same shortcoming: if the net mesh size is small enough to collect rotifers, it clogs rapidly with algae and debris and must be cleaned often. This problem is less serious when larger crustaceans are collected, because mesh size is larger. Feeding fry in tanks with zooplankton-laden water delivered directly from zooplankton culture ponds proved very efficient (Opuszynski et al. 1985). If water quality in a zooplankton culture pond is not satisfactory for larvae, fish tanks can be aerated and/or supplied with clean water.

Culture of live food

Different live foods can be cultured for cyprinid larvae. Rotifers, *B. rubens*, cladocerans, *Daphnia* sp. and *Moina* sp., brine shrimp, *Artemia salina*, and nematods, *Panagrellus* sp., are among the most commonly cultured food organisms (Rottmann and Shireman 1990). Zooplankton culture procedures are more complicated and labor intensive than the management of zooplankton culture ponds. However, high population densities can be obtained in culture facilities (e.g., 500 rotifers/ml), which makes zooplankton collection easy.

Brine shrimp are widely used where dried eggs are available (Leger et al. 1986). Their advantage consists in the simplicity of hatching and decapsulating eggs (Campton 1989). Rottmann et al. (1991) compared *B. rubens, A. salina,* and *Panagrellus* sp. as a food for intensively cultured grass carp and bighead carp larvae. Although all the live foods tested gave satisfactory results, the fish that were fed rotifers were significantly larger at the end of the 3-week feeding trials.

Feeding dry diets in intensive culture systems

Feeding larvae with live food in culture facilities with controlled environmental conditions is cost-effective only for a short period of time when total fish biomass is low. The transition period from natural to artificial diets can be accelerated, when the natural diet is gradually replaced by dry feed and when both the fish and feed are kept at high densities. The Dutch procedure used in large-scale integrated tank-and-cage culture is a good example (Huisman 1981). Common carp or grass carp larvae are stocked in 120-L aquaria at densities of about 330 larvae/L and are fed with brine shrimp during the first 5-7 days. Brine shrimp are steadily replaced by pelleted trout starter. Hatchery-raised fingerlings are then transferred to cages and are fed exclusively with pellets using demand feeders.

Predation Control

Predation is a serious problem in ponds, but it is relatively simple to prevent in cages and tanks. Proper timing of biological succession and of fish stocking is of basic importance. If fish are not stocked at the proper time, they become the prey rather than the predator. The predator-prey relationship reverses rapidly, however, in the course of larval culture as adult cyclopoid copepods and some insects that prey upon larvae are eaten by older fish.

An inexpensive and effective way to control air-breathing predatory insects in ponds is to spray diesel fuel or unrefined coconut oil mixed with soap (Asia) on the water surface. They should be sprayed at a rate of 30 L/ha.

Chemical fertilization plays a role in larval protection, because predation pressure decreases as algal blooms develop (Wolny 1970b). If succession has already advanced before a pond is stocked, organophosphate insecticides can be used to kill crustaceans and aquatic insects. However, the crustacean zooplankton recovery period is relatively long after such a treatment (Opuszynski and Shireman 1988), and quantitative and qualitative changes in zooplankton and benthic communities occur, which might be disadvantageous for fish feeding (Grygierek and Wasilewska 1979).

Frogs and toads are considered as serious predators and/or competitors of larval and juvenile fish (Pillay 1990). Current methods used to control frogs include poisoning egg masses and applying selective tadpole toxicants (Kane and Johnson 1989). Mechanical methods of frog control consist of hand removal of eggs and killing adult frogs. Larval ponds sometimes are fenced to prevent the entrance of frogs (Chen 1990).

Wild fish gain access to larval ponds with incoming water; therefore, different filtering devices are used to reduce this problem (Pillay 1990). When serious bird predation exists, small ponds can be covered with nets. Different scaring devices are mostly ineffective, as birds quickly learn to ignore them.

If larvae are fed in cages or tanks with zooplankton collected from ponds, filtering the zooplankton-laden water through properly-sized net is the basic precaution against predatory invertebrates. Flow-through cages have been designed to raise larvae in zooplankton-rich ponds. Rotifers and copepod nauplii flow into the cage, while predatory zooplankton are retained by the fine net. Different cage arrangements are used to provide water flow to the cages situated in drainable and undrainable ponds (Lirski et al. 1979;

Opuszynski et al. 1985). Flow-through cages not only protect larvae against predation and enable continual delivery of live food, but also make it easy to train larvae to accept dry feed. These cages can ensure near 100% survival of larvae, providing favorable water quality and food supply exist in the pond.

STRATEGIC AND TACTICAL DECISIONS IN CYPRINID LARVAE CULTURE

The culture of cyprinid larvae is a complex procedure. It must be carefully planned and executed to give satisfactory results. All facilities, equipment, and necessary preparations should be ready before larval culture begins. The sequence of events during culture is rapid, and there is no time to improvise. During the early larval stage, even a short exposure to adverse conditions can result in high mortality or slow growth. Larvae are usually available for only a short period during the year; therefore, a failure in larval culture may mean that a year's effort is lost.

The choice of the best larval culture strategy depends on local climatic, technical, and socio-economic conditions. The climate determines whether water should be heated and when culture should be started. Different culture strategies can be used, depending on the technology and facilities available. Pond culture is the simplest strategy from a technological standpoint, whereas tank culture, using recirculating water with fully controlled environmental conditions, is complicated.

The management strategy selected depends on the goal and scale of the culture operation. An accessory operation which provides fingerlings for a single fish farm will be smaller than one that specializes in fingerling production. The size of the larval culture operation is an important factor in management decisions, because not all culture practices can be easily increased when production increases. Currently, the largest larval culture operations raise larvae in ponds. The choice of proper strategies and tactics begins with the selection of ponds, cages, or tanks, and with consideration of water quality, food, and predation that must be controlled.

Integrated methods of culture are worthy of special recommendation. In integrated culture, the larvae are transferred as they grow from more to less controlled environments, for example, from indoor tanks to cages and/or ponds. The rationale for using these methods consists of a rapid increase of the resistance to environmental factors along with larval growth. This is especially true concerning oxygen deficiency, starvation, live food accessibility, dry feed utilization, and invertebrate predation control. Even a relatively small increase in fish weight (5-6 mg) may substantially improve survival in less controlled environmental conditions.

ACKNOWLEDGMENTS

The senior author dedicates this paper to his former co-workers from the Inland Fisheries Institute, Poland: Andrzej Lirski, Leszek Myszkowski, Grazyna Okoniewska, Barbara Onoszkiewicz, Teresa Pawlak, Malgorzata Szlaminska, Jacek Wolnicki, Michal Wozniewski, and Jadwiga Zbrojkiewicz. Journal Series No. R-02041.

REFERENCES

Alami-Durante, H., N. Charlon, A. Escaffre, and P. Bergot. 1991. Supplementation of artificial diets for common carp (*Cyprinus carpio* L.) larvae. Aquaculture 93:167-175.

Aleksandrijskaia, A. 1978. Contemporary practices in pond fertilization. Rybovodstvo i Rybolovstvo 6:15-16. (in Russian)

Anderson, R.O. 1993. Apparent problems and potential solutions for production of fingerling striped bass, *Morone saxatilis*. Journal of Applied Aquaculture 2(3/4): 101-118.

Antalfi, A., and I. Tolg. 1975. The Herbivorous Fishes. Panstwowe Wydawnictwo Rolnicze i Lesne, Warsaw. (in Polish)

Ashkierov, T.A. 1975. Survival and oxygen uptake by juvenile common carp under different conditions. Hidrobiologicheski Zhurnal 6:102-105. (in Russian)

Baranova, V.P. 1974. Dependence of the carp larval growth rate on the conditions of rearing. Izvestiya Gosudarstvennogo nauchno-issledovatelskogo instituta ozernogo i rechnogo rybnogo khozyaystva 92:66-78. (in Russian)

Bryant, P.L., and A.J. Matty. 1981. Adaptation of carp (*Cyprinus carpio*) larvae to artificial diets. I. Optimum feeding rate and adaptation age for a commercial diet. Aquaculture 23:275-286.

Campton, D.E. 1989. A simple procedure for decapsulating and hatching brine shrimp. Progressive Fish-Culturist 51:176-179.

Charlon, N., and P. Bergot. 1984. Rearing system for feeding fish larvae on dry diets. Trial with carp (*Cyprinus carpio* L.) larvae. Aquaculture 41:1-9.

Charlon, N., H. Durante, A.M. Escaffre, and P. Bergot. 1986. Alimentation artificielle des larves de carpe (*Cyprinus carpio* L.). Aquaculture 54:83-88.

Chen, L.-C. 1990. Aquaculture in Taiwan. Fishing News Books, Oxford.

Ciborowska, J. 1972. Food of Asiatic herbivorous fish (*Ctenopharyngodon idella* Val., *Hypophthalmichthys molitrix* Val., *Aristichthys nobilis* Rich.) raised together with common carp in fry ponds. Roczniki Nauk rolniczych H-94, 2:41-58. (in Polish; English summary)

Culver, D.A., S.P. Maden, and J. Qin. 1993. Percid pond production techniques: Timing, enrichment, and stocking density manipulation. Journal of Applied Aquaculture 2(3/4): 9-31.

Dabrowski, K. 1984. Influence of initial weight during the change from live to compound feed on the survival and growth of four cyprinids. Aquaculture 40:29-40.

Dabrowski, K., and R. Bardega. 1984. Mouth size and predicted food size preferences of larvae of three cyprinid fish species. Aquaculture 40: 41-46.

Dabrowski, K., and D. Culver. 1991. The physiology of larval fish: digestive tract and formulation of starter diets. Aquaculture Magazine 17(2): 49-61.

Dabrowski, K., and P. Poczyczynski. 1988. Comparative experiments on starter diets for grass carp and common carp. Aquaculture 69:317-332.

FAO (Food and Agriculture Organization of the United Nations). 1983. Freshwater aquaculture development in China. FAO Fisheries Technical Paper 215, Food and Agriculture Organization of the United Nations, Rome.

FAO (Food and Agriculture Organization of the United Nations). 1991. Fishery Statistics, Catches and Landing. FAO Yearbook 1989, Food and Agriculture Organization of the United Nations, Rome.

Fijan, N. 1967. Problems in carp pond fertilization. FAO Fisheries Report 44:114-123.

Grygierek, E., and B. Wasilewska. 1979. Regulation of fish pond biocoenosis. Pages 317-333 *in* E. Styczynska-Jurewicz, T. Backiel, E. Jaspers, and G. Persoone, eds. Cultivation of Fish Fry and Its Live Food. European Mariculture Society Special Publication 4, Bredene, Belgium.

Horoszewicz, L. 1974. Survival of starving common carp larvae at different water temperature. Roczniki Nauk rolniczych H-96, 3:45-53. (in Polish; English summary)

Huisman, E.A. 1981. Integration of hatchery, cage, and pond culture of common carp (*Cyprinus carpio* L.) and grass carp (*Ctenopharyngodon idella* Val.) in the Netherlands. Pages 266-273 *in* L.J. Allen and E.C. Kinney, eds. Proceedings of the Bio-Engineering Symposium for Fish Culture. Publication 1, Fish Culture Section of the American Fisheries Society, Bethesda, Maryland.

Ito, J., R. Suzuki. 1977. Feeding habits of a *Cyprinidae* loach fry in the early stages. Bulletin of Freshwater Fisheries Research Laboratory 27:85-94.

Januszko M., T. Bednarz, M. Broda, and E. Grzelewska. 1977. Fertilization of enclosed parts of ponds with nitrogen and phosphorus. IV. Phytoplankton. Roczniki Nauk rolniczych H-98, 1:75-104. (in Polish; English summary)

Jobling, M. 1981. Temperature tolerance and the final preferendum-rapid methods for the assessment of optimum growth temperatures. Journal of Fish Biology 19:439-455.

Jowko, G., M. Korwin-Kossakowski, and B. Jezierska. 1981. The effect of selected environmental factors on carp (*Cyprinus carpio* L.) larvae. Roczniki Nauk rolniczych 99-H:25-37. (in Polish; English summary)

Kamler, E., M. Szlaminska, A. Przybyl, B. Barska, M. Jakubas, M. Kuczynski, and K. Raciborski. 1990. Developmental response of carp, *Cyprinus carpio*, larvae fed different foods or starved. Environmental Biology of Fishes 29:303-313.

Kane, A.S., and D.L. Johnson. 1989. Use of TFM (3-Trifluoromethyl-4-Nitrophenol) to selectively control frog larvae in fish production ponds. Progressive Fish-Culturist 51:207-213.

Kapitonova, I.G. 1977. Estimation of the optimal water temperature for large-scale culture of larval carp. Rybnoe Khozyaystvo, Moskow 11:60-61. (in Russian)

Khadka, R.B., and T.R. Rao. 1986. Prey size selection by common carp (*Cyprinus carpio* var. *communis*) larvae in relation to age and prey density. Aquaculture 54:89-96.

Klastorin, L.B., and V.A. Titov. 1975. Susceptibility to oxygen deficiency in the artificially propagated fish species. Rybnoe Khozijaistvo, Moskow 1:29. (in Russian)

Knud-Hansen, C.F., T.R. Batterson, C.D. McNabb, I.S. Harahat, K. Sumantadin and H. M. Eidman. 1991. Nitrogen input, primary productivity and fish yield in fertilized freshwater ponds in Indonesia. Aquaculture 94:49-63.

Kornenko, G.C. 1971. The importance of ciliata as a food for larval Asiatic herbivorous fish. Voprosy Ikhtiology 2:303-309. (in Russian)

Krebs, J.R. 1978. Optimal foraging: decision rules for predators. Pages 23-63 *in* J.R. Krebs and N.B. Davies, eds. Behavioral Ecology: an Evolutionary Approach. Blackwell Scientific Publications, Oxford.

Kuznetshova, J. 1958. Respiration in the early larval stage of bream, carp, and perch-pike. Pages 346-358 *in* Trudy Soveshthania po Fiziologii Ryb, Akademia Nauk USSR, Moscow. (in Russian)

Lannan, J.E., R.O. Smitherman, and G. Tchobanoglous. 1986. Principles and Practices of Pond Aquaculture. Oregon State University Press, Corvallis, Oregon.

Leger, P., D.A. Bengtson, K.L. Simpson, and P. Sorgeloos. 1986. The use and nutritional value of Artemia as food source. Oceanography and Marine Biology. An Annual Review 24:521-623.

Lewkowicz, M., and S. Lewkowicz. 1976. Organic and inorganic nutrient enrichment and the living conditions of carp fry in first rearing ponds. Physico-chemical factors and the zooplankton. Acta Hydrobiologica 3:235-257.

Lincoln, E.P., T.W. Hall, and B. Koopman. 1983. Zooplankton control in mass algae cultures. Aquaculture 32:331-337.

Lirski, A., and K. Opuszynski. 1988a. Lower lethal temperatures for carp (*Cyprinus carpio* L.) and the phytophagous fishes (*Ctenopharyngodon idella* Val., *Hypophthalmichthys molitrix* Val., *Aristichthys nobilis* Rich.) in the first period of life. Roczniki Nauk rolniczych H-101, 4:11-29. (in Polish; English summary)

Lirski, A., and K. Opuszynski. 1988b. Upper lethal temperatures for carp (*Cyprinus carpio* L.) and the phytophagous fishes (*Ctenopharyngodon idella* Val., *Hypophthalmichthys molitrix* Val., *Aristichthys nobilis* Rich.) in the first period of life. Roczniki Nauk rolniczych H-101, 4:31-49. (in Polish; English summary)

Lirski A., B. Onoszkiewicz, K. Opuszynski, and M. Wozniewski. 1977. Rearing of silver carp larvae in the cages with controlled water temperature. Gospodarka rybna 2:7-10. (in Polish)

Lirski, A., B. Onoszkiewicz, K. Opuszynski, and M. Wozniewski. 1979. Rearing of cyprinid larvae in new type flow-through cages placed in carp ponds. Polskie Archiwum Hydrobiologii 26:545-549.

Littak, A., M. Wozniewski, and K. Opuszynski. 1980. Early rearing of common carp larvae under controlled environmental conditions. Inland Fisheries Institute, Olsztyn 129:1-32. (in Polish)

Lupatsheva, L.I. 1967. Feeding of grass carp on the early stages of development. Rybnoe Khozaistvo, Kiev 3:102-104. (in Russian)

Lupatsheva, L.I. 1970. Feeding relations of silver carp and grass carp larvae in the pond polyculture. Rybnoe Khozaistvo, Kiev 11:34-37. (in Russian)

Marcel, J. 1990. Aquaculture in China. Pages 946-962 *in* G. Barnabe, ed. Aquaculture, vol. 2. Ellis Horwood, New York, New York.

Okoniewska, G., and K. Opuszynski. 1988. Rearing of grass carp larvae (*Ctenopharyngodon idella* Val.) in ponds receiving heated effluents. IV. Food. Roczniki Nauk rolniczych H-101, 4:161-175. (in Polish; English summary)

Okoniewska, G., and J. Wolnicki. 1988. Food preference of common carp (*Cyprinus carpio* L.) and grass carp (*Ctenopharyngodon idella* Val.) after starvation. Roczniki Nauk rolniczych H-101, 4:71-83. (in Polish; English summary)

Okoniewska, G., W. Piotrowska-Opuszynska, and K. Opuszynski. 1988. Rearing of grass carp larvae (*Ctenopharyngodon idella* Val.) in ponds receiving heated effluents. I. Scheme of the experiment, temperature, water flow in the ponds and fish production. Roczniki Nauk rolniczych H-101, 4:111-127. (in Polish; English summary)

Olah, J., and J. Farkas. 1978. Effect of temperature, pH, antibiotics, formalin and malachite green on the growth and survival of *Saprolegnia* and *Achlya* parasitics on fish. Aquaculture 3:273-288.

Opuszynski, K. 1969. Production of plant-feeding fish (*Ctenopharyngodon idella* Val., and *Hypophthalmichthys molitrix* Val.) in carp ponds. Roczniki Nauk rolniczych H-91:219-309. (in Polish; English summary)

Opuszynski, K. 1979. Silver carp, *Hypophthalmichthys molitrix* Val., in carp ponds. II. Rearing of fry. Ekologia polska 27:93-116.

Opuszynski, K. 1987. Fresh-water pond ecosystems managed under a moderate European climate. Pages 63-91 *in* R.G. Michael, ed. Managed Aquatic Ecosystems. Elsevier Science Publishers, B.V., Amsterdam.

Opuszynski, K. 1989. The herbivorous fish production in Poland. II. Rearing of fry. Gospodarka rybna 3:10-12. (in Polish)

Opuszynski, K., and Z. Okoniewska. 1969. Survival and changes in fat and protein content of *Ctenopharyngodon idella* Val., *Hypophthalmichthys molitrix* Val., and *Cyprinus carpio* L. during wintering in ponds. Roczniki Nauk rolniczych H-4:657-670. (in Polish; English summary)

Opuszynski K., and J.V. Shireman. 1988. Pond environmental manipulation to stimulate rotifers for larval fish rearing. Roczniki Nauk rolniczych H-101, 4:183-195.

Opuszynski, K., J.V. Shireman, F.J. Aldridge, and R. Rottmann. 1985. Intensive culture of grass carp and hybrid grass carp larvae. Journal of Fish Biology 5:563-573.

Opuszynski, K., A. Lirski, L. Myszkowski, and J. Wolnicki. 1989a. Upper lethal and rearing temperatures for juvenile common carp, *Cyprinus carpio* L., and silver carp, *Hypophthalmichthys molitrix* Val. Aquaculture and Fisheries Management 20:287-294.

Opuszynski K., L. Myszkowski, G. Okoniewska, W. Opuszynska, M. Szlaminska, J. Wolnicki, and M. Wozniewski. 1989b. Rearing of common carp, grass carp, silver carp, and bighead carp larvae using zooplankton and different dry food. Polskie Archivum Hydrobiologii 2:217-230.

Ostroumova, I.N., and V.I. Turestskij. 1981. Manual for Feeding of Carp Larvae and Early Fry with Starter Ehkvizo. GosNIORKh, Leningrad. (in Russian)

Ostroumova, I.N., V.I. Turestskij, D.I. Ivanov, and M.A. Dementyeva. 1980. Balanced starter feed for rearing of common carp larvae under heated-water conditions. Rybnoe Khozyaystvo, Moscow 6:41-44. (in Russian)

Panov, D., V. Vinogradov, and L. Chromov. 1969. Larval fish rearing. Rybovodstvo i Rybolovstvo 1:8-9. (in Russian)

Penaz, M., M. Prokes, J. Kouril, and J. Hamackova. 1983. Early development of the carp, *Cyprinus carpio* L. Acta scientiarium naturalium Academiae Scientiarium Bohemoslovacae, Brno 17:1-35.

Pillay, T.V.R. 1990. Aquaculture: Principles and Practices. Fishing News Books, Oxford.

Piotrowska-Opuszynska, W. 1984. The influence of nitrogen fertilizers on physico-chemical conditions in nursery ponds. Roczniki Nauk rolniczych H-100, 4:111-132. (in Polish; English summary)

Prost, M. 1980. Fish Diseases. PWRiL, Warsaw. (in Polish)

Rottmann, R.W., and J.V. Shireman. 1990. Hatchery Manual for Grass Carp and Other Riverine Cyprinids. Bulletin 244, Florida Cooperative Extension Service and Institute of Food and Agricultural Sciences, University of Florida, Gainesville, Florida.

Rottmann, R.W., J.V. Shireman, and E.P. Lincoln. 1991. Comparison of three live foods and two dry diets for intensive culture of grass carp and bighead carp larvae. Aquaculture 96:269-280.

Savin, G.I., R.A. Savina, and A.M. Bagrov. 1973. How to increase the survival rate of juvenile Asiatic herbivorous fish. Rybovodstvo i Rybolovstvo 2:8-9. (in Russian)

Schlumpberger, W. 1980. Zur Kombination der Rinnen und Gasekafiganfzucht von Silberkarpfenbrut (*Hypophthalmichthys molitrix*). Zeitschrift für die Binnenfischerei der DDR 27:172-174.

Schluter, M., and J. Groeneweg. 1981. Mass production of freshwater rotifers on liquid wastes. I. The influence of some environmental factors on population growth of *Brachionus rubens* Ehrenberg 1938. Aquaculture 25:17-24.

Schroeder, G.L., G. Wohlfarth, A. Alkon, A. Halevy, and H. Krueger. 1990. The dominance of algal-based food webs in fish ponds receiving chemical fertilizers plus organic manures. Aquaculture 86:219-229.

Shirota, A. 1970. Studies on the mouth size of fish larvae. Bulletin of the Japanese Society of Scientific Fisheries 36:353-368.

Skhorbatov, G.L. 1963. Acclimatization of coregonid fish in Kharkovsk district. Trudy Soveshthania Gidrobiologicheskovo Obshthestva, Moscow 13:242-254. (in Russian)

Sukhanova, E.R. 1968. Influence of cyclopoids on survival of silver carp fry. Pages 217-220 *in* G.V. Nikolskij, ed. New Studies on Ecology and Culture of Asiatic Herbivorous Fish. Izdatelstvo Nauka, Moscow. (in Russian)

Suzuki, R., M. Yamaguchi, and K. Ishikawa. 1977. Differences in growth rate in two races of common carp in various temperatures. Bulletin of the Freshwater Fisheries Research Laboratory, Tokyo, 27:21-26.

Tamas, G. 1979. Rearing of common carp fry and mass cultivation of its food organisms in ponds. Pages 281-288 *in* E. Styczynska-Jurewicz, T. Backiel, E. Jaspers, and G. Persoone, eds. Cultivation of Fish Fry and Its Live Food. European Mariculture Society Special Publication 4, Bredene, Belgium.

Tatarko, K.J. 1970. Sensitivity of common carp to high temperature at the first postembrionic stages of development. Gidrobiologicheski Zhurnal 6:102-105. (in Russian)

Trzoch-Szalkiewicz, G. 1971. Food consumed by carp fry as an element of utilization of pond production. Polskie Archiwum Hydrobiologii 2:157-165.

Trzoch-Szalkiewicz, G. 1972. Influence of stock density and pond fertilization on carp feeding habits in fry ponds. Roczniki Nauk rolniczych H-94, 2:127-141. (in Polish; English summary)

Van der Wind, J.J. 1979. Techniques of rearing phytophagous fishes. FAO Fisheries Report 44:227-232.

Vanhaecke, P., and P. Sorgeloos. 1983. International study on *Artemia*. XXX. Bio-economic evaluation of the nutritional value for carp (*Cyprinus carpio* L.) larvae of nine *Artemia* strains. Aquaculture 32:285-293.

Vinberg, G.G., and V.P. Lakhnovitch. 1965. Pond Fertilization. Pischtchevaia Promyslennost, Moscow. (in Russian)

Vovk P.S. 1976. Biology of Asiatic Herbivorous Fish and Their Use in Water Reservoirs of Ukraina. Naukova Dumka, Kiev. (in Russian)

Wolnicki, J., and K. Opuszynski. 1988. "Point of no return" in carp (*Cyprinus carpio* L.) and the phytophagous fish (*Ctenopharyngodon idella* Val., *Hypophthalmichthys molitrix* Val., *Aristichthys nobilis* Rich.) larvae. Roczniki Nauk rolniczych H-101, 4:61-69. (in Polish; English summary)

Wolny, P. 1970a. The effect of intensification measures on growth, survival and production of Asiatic herbivorous fish. Roczniki Nauk rolniczych H-92:97-119. (in Polish; English summary)

Wolny, P. 1970b. Results of the six-year study on the effectiveness of nursery pond fertilization. Roczniki Nauk rolniczych 91-H, 4:565-588. (in Polish; English summary)

Wozniewski, M., and L. Myszkowski. 1987. Oxygen conditions during common carp larval rearing at high water temperature. Gospodarka rybna 6:14-15. (in Polish)

Wozniewski, M., and K. Opuszynski. 1988. Threshold oxygen content for juvenile stages of the cyprinids (*Ctenopharyngodon idella* Val., *Hypophthalmichthys molitrix* Val., *Aristichthys nobilis* Rich., *Cyprinus carpio* L.). Roczniki Nauk rolniczych H-101, 4:51-59. (in Polish; English summary)

Wozniewski, M., A. Littak, and K. Opuszynski. 1980. Production of the silver carp stocking material. Part II. Rearing of larvae under controlled conditions. Inland Fisheries Institute, Olsztyn, 128:1-40. (in Polish)

Liming and Fertilization
of Brackishwater Shrimp Ponds

Claude E. Boyd
Harry V. Daniels

ABSTRACT. Agricultural limestone and burnt lime are applied either to the water during shrimp production or to pond bottoms between shrimp crops. However, unless the total alkalinity and total hardness of pond water is below 50 mg/L as equivalent $CaCO_3$ or the pond soils are acidic (pH < 7), liming is of little or no value. The use of burnt lime should be avoided because this material can cause high pH in water and soil. Chemical fertilizers or manures are used to fertilize brackishwater ponds. Fertilization programs for brackishwater ponds usually require more nitrogen (N) than those for freshwater ponds. Phosphorus (P) fertilization is important both in brackishwater and freshwater ponds. Because water is exchanged often in brackishwater ponds, fertilizer should be applied in small doses and at frequent intervals. Most managers of brackishwater ponds prefer a large proportion of diatoms in the phytoplankton community. An N:P application ratio of 20:1 in ponds favors diatoms; in fiberglass tanks with water of low silica concentration, fertilization with silica encouraged an abundance of diatoms.

Claude E. Boyd, Department of Fisheries and Allied Aquacultures, Alabama Agricultural Experiment Station, Auburn University, Alabama 36849, USA.

Harry V. Daniels, Continental Grain Company, Desarrollo Industrial Bioacuatico, S. A., 303 Madison Avenue, 9th Floor, New York, NY 10017, USA.

[Haworth co-indexing entry note]: "Liming and Fertilization of Brackishwater Shrimp Ponds." Boyd, Claude E., and Harry V. Daniels. Co-published simultaneously in the *Journal of Applied Aquaculture* (The Haworth Press, Inc.) Vol. 2, No. 3/4, 1993, pp. 221-234; and: *Strategies and Tactics for Management of Fertilized Hatchery Ponds* (ed: Richard O. Anderson and Douglas Tave) The Haworth Press, Inc., 1993, pp. 221-234. Multiple copies of this article/chapter may be purchased from The Haworth Document Delivery Center [1-800-3-HAWORTH; 9:00 a.m. - 5:00 p.m. (EST)].

INTRODUCTION

Liming is a common practice at all levels of shrimp farming, but it is probably used most widely in intensive shrimp farming. Fertilizer is used most commonly in extensive and semi-intensive shrimp farming where little or no feed is applied to ponds. Although liming and fertilization are common techniques in shrimp production, there has been little research on their use, and current practices vary widely. Shrimp pond managers often do not have data on the properties of liming materials they use in ponds or understand the effects of liming. Properties of fertilizers are well known, but many different fertilization techniques are used. The purpose of this report is to discuss "state-of-the-art" practices of liming and fertilizing shrimp ponds.

LIMING

Liming Materials

Three basic products are used for liming shrimp ponds: pulverized limestone, burnt limestone, and hydrated lime. Natural deposits of limestone are comprised of calcium carbonate or a mixture of calcium and magnesium carbonates. Limestone rock may be mined and finely crushed to particle sizes of 10 mesh to 60 mesh (1.70 mm to < 0.24 mm) with a rock crusher. Pulverized limestone is called "agricultural limestone." Burnt lime is prepared by heating limestone in a furnace. The limestone decomposes to yield calcium oxide:

$$CaCO_3 \xrightarrow{\Delta} CaO + CO_2$$

The limestone used for preparing burnt lime may not be pure calcium carbonate, so burnt lime may also contain some magnesium oxide. Hydrated lime (calcium hydroxide) is prepared by treating burnt lime with water:

$$CaO + H_2O \rightarrow Ca(OH)_2$$

This product also may contain some magnesium hydroxide.

Where production facilities are crude, complete decomposition of limestone may not be achieved, and burnt lime may contain some calcium carbonate. Thus, some batches of calcium oxide and calcium hydroxide may contain significant quantities of calcium carbonate.

Calcium oxide is called "quick lime" or "unslaked lime," and calcium hydroxide is known as "hydrated lime" or "slaked lime" in the United States. In Spanish speaking countries, agricultural limestone is termed *"cal agricola,"* burnt lime is called *"cal viva,"* and hydrated lime is known as *"cal hidratada."*

In addition to materials prepared from limestone rock, soft marl deposits, sea shells, and wood ash sometimes are used as liming materials. These substances contain calcium carbonate (marl and sea shells) or oxides and hydroxides of sodium, potassium, calcium, and magnesium (wood ash). Such substances normally contain more impurities than limestone rock and are less effective than liming materials made from limestone.

Liming materials react vigorously to neutralize acidity:

$$CaO + 2H^+ \leftrightarrows Ca^{2+} + 2H_2O$$

$$Ca(OH)_2 + 2H^+ \leftrightarrows Ca^{2+} + 2H_2O$$

$$CaCO_3 + 2H^+ \leftrightarrows Ca^{2+} + CO_2 + H_2O$$

They also combine with carbon dioxide to form bicarbonate:

$$CaO + 2CO_2 + H_2O \leftrightarrows Ca^{2+} + 2HCO_3^-$$

$$Ca(OH)_2 + 2CO_2 \leftrightarrows Ca^{2+} + 2HCO_3^-$$

$$CaCO_3 + CO_2 + H_2O \leftrightarrows Ca^{2+} + 2HCO_3^-$$

When added to a pond, liming materials will neutralize acidity and increase the pH of water and soil. They will increase calcium and magnesium ion concentrations in the water and thus raise the total hardness. They will remove carbon dioxide from the water. Where ponds are highly acidic, part of the anionic component of the liming material (oxide, hydroxide, or carbonate) will be converted to water or water and carbon dioxide in neutralizing acidity. However, any excess oxide, hydroxide, or carbonate will react with carbon dioxide to form bicarbonate ion and raise total alkalinity.

All liming materials are sparingly soluble in water. The rate of dissolution increases with decreasing particle size. However, it is normally difficult to increase total alkalinity and total hardness above 50 to 60 mg/L as $CaCO_3$ equivalent with liming materials, because this is the solubility limit of liming materials at carbon dioxide concentrations normally found in pond water. Under very acidic conditions or at high carbon dioxide concentrations, it is possible to dissolve a greater amount of limestone in water. Of course, in water with pH < 5, the increase in total hardness following liming will be greater than the increase in total alkalinity, because bicarbonate ion is expended in neutralizing acidity, but the calcium and magnesium ions remain in the water.

Liming materials may be classified according to neutralizing value and fineness. The neutralizing value compares the ability of a liming material to neutralize acidity to that of pure calcium carbonate. A neutralizing value of 100% is assigned to pure calcium carbonate. Pure calcium oxide has a neutralizing value of 179%, while that of pure calcium hydroxide is 136%. Commercial liming materials are not pure, and neutralizing values differ depending upon the type and quality of the product. Twenty samples of shrimp pond liming materials from Ecuador, Honduras, Thailand, and the United States were analyzed for neutralizing value by the procedure of Boyd (1990); values ranged from 80.7 to 167.9%.

The fineness of liming materials refers to the size of its constituent particles. Particles which pass a 60-mesh screen are considered most effective because they dissolve fastest. Particles too large to pass a 10-mesh screen dissolve so slowly that they are of little use in pond liming. A scheme for evaluating the fineness of liming materials for ponds was developed by Boyd and Hollerman (1982). In this scheme, a fineness value of 100% is assigned to a material in which all particles will pass a 60-mesh screen. The fineness value decreases with increasing particle size. A material having no particles which will pass a 10-mesh screen has a fineness value of 3.6%. The 20 samples of shrimp pond liming materials mentioned above had fineness values of 60 to 100%. Application rates for agricultural limestone mentioned here are for fine grain material, most of which will pass a 60-mesh screen. Application rates should be doubled or even tripled for coarse grain material.

Although neutralizing value and fineness analyses are not difficult to make, most nations have no regulations regarding the quality and labelling of commercial liming materials. Therefore, the composition and properties of liming materials used in shrimp ponds often are unknown. One usually can distinguish agricultural limestone from burnt lime or hydrated lime by the pH of a slurry of 10 to 20 parts distilled water and 1 part liming material. Agricultural limestone will not raise the pH above 9.5 or 10, while the pH of burnt lime or hydrated lime normally will exceed 12. Burnt lime and hydrated lime do not give off heat when dissolved in water, as is commonly reported by shrimp farmers. Caution must be exercised when handling burnt and hydrated lime, for they can damage skin and eyes.

Liming During Crops

Brackish waters for use in shrimp ponds usually have total alkalinity values between 50 and 150 mg/L as $CaCO_3$. Such waters are well-buffered against pH change, and even when alkalinity of pond water is neutralized by acidity sources in the pond, the alkalinity may be replenished through water exchange. There is some evidence that shrimp grow best when total alkalinity is above 100 mg/L as $CaCO_3$. However, because of the low solubility of liming materials, it is difficult to achieve total alkalinity values above 50 to 60 mg/L as $CaCO_3$ following their use. It is not advisable to apply liming materials to ponds during the grow-out period unless total alkalinity values are below 40 or 50 mg/L as $CaCO_3$. Even then, the effect of liming will be short-lived if water exchange rates are high. Methods for estimating liming rates for brackishwater ponds with water exchange have not been developed, so it is probably best to apply agricultural limestone in increments of 500 to 1,000 kg/ha at 2-week intervals until the desired total alkalinity concentration is reached. Afterwards, additional increments can be applied when total alkalinity falls below the target concentration. To avoid dangerously high pH, burnt lime should not be used in ponds during the grow-out period.

Some shrimp farmers apply 2,000 kg/ha of agricultural limestone to ponds just before harvesting shrimp. They claim that the added calcium "hardens" shrimp so that they are in better condition at harvest. There are no data to support the validity of this procedure. Some workers also have incorporated liming materials in mixed fertilizers, with the idea that the added calcium is beneficial when shrimp molt. The amount of calcium added to the water in this manner is insignificant in brackishwater ponds. Commercial, mixed fertilizers must contain a filler in order to obtain desired percentages of nitrogen and phosphorus in the bulk material. Agricultural limestone is an excellent filler for mixed fertilizers.

Liming Between Crops

In intensive shrimp farming in particular, pond managers want to treat pond bottoms between crops to lower soil organic matter concentrations as much as possible. To accomplish this, managers often dry pond bottoms, plow them to a depth of 10 to 15 cm, and apply liming materials. Drying and plowing provide better contact with air to enhance microbial activity. Microbial activity in soil is greatest at a pH of 7 to 8. Therefore, applying liming materials to acidic soils should foster greater rates of organic matter decomposition in dry pond soils. Of course, burnt and hydrated lime can raise the pH of soils so high that microbial organisms are killed and decomposition is temporarily stopped or greatly retarded. Only agricultural limestone should be applied to dry pond bottoms as an aid to organic matter decomposition.

A method for determining the amount of agricultural limestone needed to raise soil pH above 7 was developed by Pillai and Boyd (1985). Where it is not feasible to use this procedure, the following technique is recommended: measure the pH of a slurry made by mixing soil and water in a weight-to-volume ratio of 1:1. Use the pH to obtain the liming rate:

pH	Agricultural limestone (kg/ha)
<5	3,000
5 to 6	2,000
6 to 7	1,000

Excessive drying of pond bottoms will result in too little moisture to dissolve agricultural limestone and for microbial activity.

In de-watered shrimp ponds, some areas in the bottom appear very black because they are anaerobic. Some managers apply burnt or hydrated lime at 0.1 to 0.2 kg/m^2 over such areas. There are no data that suggest there is any benefit to this practice.

When it is desired to sterilize pond bottoms, burnt or hydrated lime often is applied at 1,000 to 2,000 kg/ha. This is thought to raise the pH of the soil enough to kill pathogenic bacteria which can reside in soils between crops. Of course, for this treatment to be effective, the liming material must be relatively pure calcium oxide or calcium hydroxide, and it must be very evenly applied and well-mixed with the soil. Agricultural limestone and burnt or hydrated lime that contain a high percentage of un-decomposed limestone will not increase the soil pH enough to kill pathogens.

In applying liming materials to pond bottoms, it is essential to provide uniform distribution. Even when a good quality liming material is applied properly to dry pond soils, there is no reliable information available on its benefits to shrimp production. Benefits of liming shrimp ponds are simply assumed, because liming has proven effective in acidic, freshwater fish ponds by improving total alkalinity, total hardness, soil and water pH, rates of primary productivity, and fish production.

Acid-sulfate Soils

In some coastal areas, potential acid-sulfate soils are used for pond construction. Such soils contain iron pyrite, FeS_2; when exposed to air, the iron pyrite oxidizes to form sulfuric acid. Oxidation of pyrite occurs primarily in surface soils of levees and in pond bottoms which have dried between crops. Pyrite is thought to oxidize very slowly in flooded soils.

There are three different procedures for identifying potential acid-sulfate soils:

1. Dry soil for several weeks. Measure the pH of a 1:1 mixture of dry soil and distilled water. A pH below 3.5 indicates a potential acid-sulfate soil.
2. Measure the total sulfur content. A value of 0.5% or more suggests a potential acid-sulfate soil.
3. Wet a few grams of fresh soil with 30% hydrogen peroxide, H_2O_2, and mix thoroughly. Measure the pH with Universal pH

paper (pH range 0 to 6). A pH below 2.5 reveals a potential acid-sulfate soil.

The lime requirement of potential acid-sulfate soils usually ranges between 25 to 150 ton $CaCO_3$/ha. It is not feasible to add this much liming material. New ponds on potential acid-sulfate soils can be subjected to successive cycles of drying, flooding, and flushing to oxidize pyrite and remove the resulting sulfuric acid. Other procedures for mitigating potential acid-sulfate soils are:

1. Levees should have as small a surface area as possible.
2. Pond should be no deeper than necessary.
3. Levees should be covered with grass to provide a barrier between air and soil and to minimize contact between runoff and soil. Levees must be limed at 0.05 to 0.1 kg/m^2, fertilized, and irrigated to establish grass.
4. Pond should be kept full of water at all times.
5. After harvest, pond should be refilled immediately to prevent drying of pond bottom.
6. Total alkalinity should be monitored, and lime should be applied at 1,000 kg $CaCO_3$/ha when values fall below 40 or 50 mg/L as $CaCO_3$.

FERTILIZATION

There has been a tremendous amount of research on fertilization of freshwater fish ponds. This research has resulted in the development of fertilization programs for chemical fertilizers and manures, which will increase primary productivity by phytoplankton and ultimately increase fish production. Depending upon fish species and the natural fertility of water, fertilization may increase production 2- to 10-fold (Boyd 1990). Although fertilization can be cost effective, fertilization will not provide the levels of fish production possible with feeding. Shrimp farmers have had the same experience, and fertilization is seldom used alone. It is used to enhance the production of natural food organisms for shrimp in ponds where feed applications are the primary source of nutrients for shrimp. It is generally accepted by pond managers that shrimp production is better, even in ponds to which feed is applied, if there is a stable and relatively high abundance of phytoplankton. In semi-intensive shrimp farming as practiced in Ecuador and many other South and Central American countries, feeding rates normally do not exceed 25 or 30 kg/ha daily in the final weeks of the grow-out period. Therefore, during the first half of the grow-out period, fertilizers often are applied to bolster phytoplankton growth.

There has been little research on fertilization of shrimp ponds. Most shrimp farm managers feel that procedures for fertilizing shrimp ponds must take into account the following differences between brackishwater shrimp ponds and freshwater fish ponds:

1. Brackishwater ponds have a greater requirement for nitrogen than freshwater ponds, because blue-green algae are comparatively scarce in brackish water.
2. Because of water exchange, fertilizer should be applied to brackishwater ponds at more frequent intervals than to freshwater ponds.
3. An N:P ratio of 20:1 or higher favors the development of diatom blooms which often are considered desirable in shrimp ponds, but data to confirm this widely-held opinion are unavailable.
4. Phosphorus solubility is low in brackishwater ponds because of the high calcium concentrations, and fertilizers should be applied frequently.

Four fertilization programs are in common use in shrimp ponds in Ecuador, and these procedures are given because they are effective and they are similar to procedures used in many other shrimp-farming nations. Most pond managers want a plankton bloom that restricts the Secchi disk visibility to 30 to 40 cm. However, some pond managers feel that the Secchi disk visibility is not a good measure of phytoplankton abundance, because it is influenced by other kinds of turbidity. These managers make counts of total phytoplankton abundance. An abundance above 300,000 to 400,000 phytoplankton cells/ml is usually considered sufficient. Fertilizers are applied at intervals, but on a scheduled application date, the application is made only if the Secchi disk visibility is greater than desired or if the abundance of phytoplankton cells is too low. The four procedures are:

1. Treat ponds at 2- to 3-day intervals with 10 kg/ha urea and 5 kg/ha triple superphosphate.
2. Treat ponds with 10 to 20 kg/ha of urea and 0.3 to 0.6 kg/ha of triple superphosphate at intervals of 1 day to 1 week. This procedure encourages a high abundance of diatoms in the plankton, and applications are made less frequently once a satisfactory abundance of diatoms is achieved. Some managers measure soluble phosphate concentrations in pond water, and when values are above 100 μg/L as P, they apply urea only.
3. Apply 2.5 to 5 kg/ha of urea daily and 1.5 to 2.5 kg/ha of triple superphosphate every other day.
4. Apply urea at 5 to 10 kg/ha initially and make repeated applications of 2 kg/ha urea at 2- to 3-day intervals. Some farmers double the rate in nursery ponds. This procedure is followed primarily in ponds where feeding rates are fairly high, and it is desired to increase the proportions of diatoms in the phytoplankton community.

There is a growing tendency for researchers to report fertilizer application rates in terms of concentration of the element, i.e., N or P in μg/L. This practice is seldom used in commercial aquaculture. Ponds in Ecuador average about 1 m deep. Therefore, 1 kg/ha of triple superphosphate (20.1% P) is equal to 20 μg P/L, and 1 kg/ha of urea (45% N) is equal to 45 μg N/L. Treatment 2 above uses additions of 450-900 μg N/L and 60-120 μg P/L for ponds 1 m deep.

Shrimp production in fertilized ponds usually is in the range of 250 to 500 kg/ha per crop. Therefore, feed also normally is applied to ponds in Ecuador to increase production. With feeding or feeding plus fertilization, shrimp production often reaches 1,000 to 2,000 kg/ha per crop. Early in the production cycle, feeding rates are low and N and P fertilizers are used to enhance phytoplankton and benthos production. Later, when feeding rates are high enough to support good phytoplankton blooms, urea applications are made to stimulate diatom production. A diatom abundance of 20 to 30% of the total phytoplankton cells is considered adequate.

Diatom abundance is favored by high N:P ratios in fertilized, brackishwater shrimp ponds (Figure 1). These data suggest that a 15:1 or a 30:1 N:P ratio will increase the proportion of diatoms in the phytoplankton community. Shrimp farmers usually think that urea is more effective in stimulating diatoms than ammonium or nitrate fertilizer. The proportion of diatoms was generally greater in tanks treated with urea than in tanks treated with ammonium chloride or sodium nitrite (Table 1).

Manure is sometimes applied to shrimp ponds. Rates vary tremendously, but the usual rates seldom exceed 250 kg/ha of fresh manure per week. Shrimp production in ponds treated at 250 kg/ha per week of manure is roughly equal to that achieved with chemical fertilizers. Higher applications of manure can result in greater shrimp production than chemical fertilizers. Wyban et al. (1987) applied cow manure at 1,800 kg/ha/week and produced 1,860 kg/ha of shrimp. Aeration and water exchange of 20% pond volume per day was necessary to prevent oxygen depletion in the heavily-ma-

FIGURE 1. Phytoplankton composition (as percent of total phytoplankters) in eight 1-ha ponds fertilized with four different ratios of nitrogen: phosphorus (N:P as kg/ha) during May, 1988 at the Desarrollo Industrial Bioacuatico Sociedad Anonyme (DIBSA) shrimp farm Guayaquil, Ecuador. Legend: shaded = Bacillariophyceae (diatoms); white = Cyanophyceae (blue-green algae); stippled = Chlorophyceae (green algae); remainder = Dinophyceae, Euglenophyceae, and Xanthophyceae. Arrows indicate dates of fertilizer application.

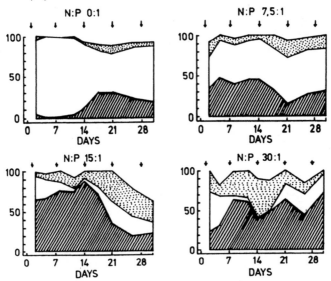

TABLE 1. Percentages of diatoms (Bacillariophyceae) in fiberglass tanks filled with brackish water and fertilized with nitrogen, phosphorus, and silicate.

N:P:Si[*] (kg/ha)	Nitrogen Source		
	Ammonium chloride	Sodium nitrate	Urea
0:3:0	1.5	0.4	16.9
0:3:30	49.0	39.7	61.8
1:1:0	3.1	2.2	10.2
1:1:30	40.0	76.9	47.6
15:1:0	7.5	0.5	23.3
15:1:30	49.9	35.3	81.0
30:1:0	0.4	31.6	22.4
30:1:30	37.7	95.5	44.2

[*] 1 kg/ha = 100 μg/l.

nured ponds. The value of supplementing manure applications with chemical fertilizers has not been demonstrated in shrimp ponds.

There is evidence that fertilization with silica can increase the abundance of diatoms in brackishwater ponds (Daniels 1989). Data in Table 1 show that silica application as sodium metasilicate to tanks filled with brackish water increased the proportions of diatoms above that achieved with nitrogen fertilization. Initial silica concentrations were below 1 mg/L in the tanks. Many waters have concentrations of silica above 1 mg/L, so these findings may not be widely applicable. Further research will be necessary to determine the general usefulness of silica fertilization, but some shrimp farmers apply sugar cane ash, which is high in silica, to ponds.

Feeding can result in much greater shrimp yields than can be achieved with chemical fertilizers or manures. Production of 4,000 to 8,000 kg/ha of shrimp per crop is common in intensive ponds in

Asia with feeding. Nutrients which enter the water in shrimp excrement and from decomposition of uneaten feed cause heavy plankton blooms, so fertilizers and manures are not normally applied to ponds with high feed inputs. However, even for intensive shrimp ponds in Asia, a few pond managers use periodic applications of urea in an attempt to increase the proportion of diatoms in the plankton community.

There is evidence that applications of nitrogen fertilizer will enhance the decomposition of organic matter in pond bottoms. When ponds are dry between crops, some farmers in Ecuador apply 24 to 48 kg/ha of urea to the bottom. Three days later they will treat with 500 kg/ha of finely-pulverized agricultural limestone. If only coarsely-pulverized limestone is available, they treat with 2,000 kg/ha. Bottom treatment with urea is probably most effective in ponds that have high concentrations of undecomposed vegetative material left from initial pond construction. Organic matter produced in ponds or applied in feed usually contains adequate nitrogen to effect its decomposition.

In semi-intensive shrimp farming, mechanical aeration is not applied, and water exchange is practiced to prevent low dissolved oxygen concentrations. Most managers exchange 5 to 15% of pond volume daily as a precautionary measure whether dissolved oxygen concentrations are low or not. Of course, this action flushes out fertilizer nutrients and phytoplankton. Because of the high cost of pumping water, some pond managers are reducing water exchange rates, and they are trying to coordinate water exchange with periods when pond water quality is impaired. Changes in water exchange regimes will alter fertilization requirements.

The need for research and on-farm testing of fertilizer and liming practices in shrimp ponds is obvious. The indiscriminate use of liming is not justified by the knowledge of effects of liming on ponds, and in many instances, liming probably does not result in tangible benefits. Although fertilization can result in increased shrimp production, there are many unknown factors, i.e., benefits of diatoms to shrimp production, effects of N:P ratios and silica fertilization on diatom growth, proper rates and frequency of application for fertilizers, relative benefits of fertilizers and feeds, etc. However, acquisition of reliable data on shrimp pond management techniques is difficult, because shrimp farming is conducted primarily in countries where there is little governmental support of aquacultural research. On-farm research conducted by shrimp producers is almost invariably designed without proper replication and control, because producers are unwilling to sacrifice short-term profits in order to obtain conclusive data.

REFERENCES

Boyd, C. E. 1990. Water Quality in Ponds for Aquaculture. Alabama Agricultural Experiment Station, Auburn University, Alabama.

Boyd, C. E., and W. D. Hollerman. 1982. Influence of particle size of agricultural limestone on pond liming. Proceedings of the Southeastern Association of Fish and Wildlife Agencies 36:196-301.

Daniels, H. V. 1989. Water Quality Studies in Brackishwater Ponds. Doctoral dissertation, Auburn University, Alabama.

Pillai, V. K., and C. E. Boyd. 1985. A simple method for calculating liming rates for fish ponds. Aquaculture 46:157-162.

Wyban, J. A., C. S. Lee, V. T. Sato, J. N. Sweeney, and W. K. Richards, Jr. 1987. Effect of stocking density on shrimp growth rates in manure-fertilized ponds. Aquaculture 61:23-32.

Water Management to Control
Clam Shrimp, *Cyzicus morsie,*
in Walleye, *Stizostedion vitreum,*
Production Ponds

James M. Czarnezki
Ernest J. Hamilton
Bruce A. Wagner

ABSTRACT. Clam shrimp, *Cyzicus morsie,* can occur in hatchery ponds in such dense numbers that they interfere with production of fish. Hatchery ponds are frequently left dry during the fall and winter and are filled in spring or early summer, simulating the vernal pools where clam shrimp naturally occur. Ponds left dry over winter and ponds that were full over winter, but were drained and dried for a period of time immediately prior to stocking, had the highest numbers of clam shrimp ($P = 0.001$). Few or no clam shrimp were collected in ponds that were full over winter and were not dried in the spring. High turbidities were observed in ponds with high numbers of clam shrimp; however, clam shrimp were not the only cause of turbidity. The largest number of walleye were produced in ponds

James M. Czarnezki, Missouri Department of Conservation, Fish and Wildlife Research Center, 1110 South College Avenue, Columbia, MO 65201, USA.

Ernest J. Hamilton, Blind Pony Hatchery and Wildlife Area, Route 2, Box 17, Sweet Springs, MO 65351, USA.

Bruce A. Wagner, Missouri Department of Conservation, Fish and Wildlife Research Center, 1110 South College Avenue, Columbia, MO 65201, USA.

[Haworth co-indexing entry note]: "Water Management to Control Clam Shrimp, *Cyzicus morsie,* in Walleye, *Stizostedion vitreum,* Production Ponds." Czarnezski, James M., Ernest J. Hamilton, and Bruce A. Wagner. Co-published simultaneously in the *Journal of Applied Aquaculture,* (The Haworth Press, Inc.) Vol. 2, No. 3/4, 1993, pp. 235-242; and: *Strategies and Tactics for Management of Fertilized Hatchery Ponds* (ed: Richard O. Anderson and Douglas Tave) The Haworth Press, Inc., 1993, pp. 235-242. Multiple copies of this article/chapter may be purchased from The Haworth Document Delivery Center [1-800-3-HA-WORTH; 9:00 a.m. - 5:00 p.m. (EST)].

235

which contained few or no clam shrimp. Clam shrimp were controlled by preventing hatchery pond substrate from drying during winter and early spring; however, the absence of drying may cause other problems that interfere with fish production.

INTRODUCTION

Clam shrimp, *Cyzicus morsie,* (Eubranchiopoda: Conchostraca) can interfere with fish production in warmwater hatchery ponds. When clam shrimp populations reach nuisance levels, they cause excessive turbidity which interferes with spawning, fry collection, and photosynthesis (McCraren and Phillips 1977). Large numbers of clam shrimp greatly reduce the desirable zooplankton upon which fish feed (McCraren et al. 1977). Clam shrimp can impede fish harvest by clogging screens and suspending sediment, which fills the kettles in hatchery ponds.

Eubranchiopoda (tadpole shrimp, fairy shrimp, and clam shrimp) characteristically inhabit temporary ponds and pools, and are most plentiful during spring and early summer (Pennak 1953). Drying and freezing were reported as a stimulus for hatching Eubranchiopoda eggs (Weaver 1943). The management of many warmwater hatchery ponds simulates the temporary pools in which clam shrimp naturally thrive. After fish harvest in summer, hatchery ponds are dry during the fall and winter and are filled the following spring or early summer, similar to the cycle in the vernal pools where clam shrimp naturally occur. Dexter and McCarraher (1967) suggested interrupting the wet-dry cycle by not allowing ponds to dry following fish harvest, in order to control clam shrimp.

Clam shrimp can be controlled with chemicals such as Dylox[1] (McCraren et al. 1977); however, these chemicals also kill Cladocera (McCraren and Phillips 1977), a highly desirable and important food item of most species raised in Missouri Department of Conservation hatchery ponds. Liners can be placed in ponds, which create a barrier between the water and pond substrate. This is effective, but it is very expensive. Biological controls, such as redear sunfish, *Lepomis microlophus,* have potential but have not been evaluated. The objective of

---

1. Use of trade or brand name does not imply endorsement.

this study was to develop a method to control clam shrimp in hatchery ponds by means of water management.

MATERIALS AND METHODS

Hatchery Ponds

This study was conducted at Blind Pony Hatchery near Sweet Springs, Missouri. Clam shrimp numbers were enumerated in earthen ponds using four different water management regimens during 1991. Each treatment was applied in two 0.4-ha ponds:

Treatment 1 – Ponds were drained for fish harvest and left empty from June until April, when they were filled and stocked with fry.

Treatment 2 – Ponds were drained for fish harvest and refilled immediately. The ponds were left full from June to February. They were drained in March to allow sediment to dry and refilled in April and stocked with fry.

Treatment 3 – Ponds were drained for fish harvest in late May, left empty during June, filled in July, and kept full over winter. The ponds were stocked with fry in April.

Treatment 4 – Ponds were drained for fish harvest in late May and refilled immediately. The ponds were kept full over winter and stocked with fry in April.

Clam shrimp numbers were monitored using funnel traps similar to those used by Whiteside (1974). The funnel trap consisted of a clear plastic (polymethyl-pentene) funnel (20.3-cm diameter) fitted to a 1-L glass bottle. The funnel was inverted and held 5-7 cm above the sediment by three metal legs attached to the funnel. Five funnel traps were placed in each pond at a depth of 1-1.25 m for 24 hours. Clam shrimp were collected in the funnel traps as they moved vertically from the substrate. The ponds were sampled weekly for 5 weeks beginning in early May. Samples were preserved in alcohol and sorted, and the clam shrimp were counted. A

Friedman test (Conover 1980) was used to determine if the numbers of clam shrimp collected in the eight ponds (two ponds per treatment) were significantly different ($P = 0.001$).

Surface water samples were collected weekly, for 5 weeks, in collapsible plastic, 1-L containers, for turbidity analysis. Turbidity measurements were made with a Hach 2100 Turbidimeter and expressed in Jackson Turbidity Units (JTU).

The ponds were stocked at a rate of 300,000 walleye fry/ha. Ponds were drained and the walleye fingerlings harvested during the last week of May or the first week of June.

Laboratory

During 1990, wet substrate was collected from two ponds, which contained large numbers of clam shrimp, immediately after draining and fish harvest. A sample of the upper 1-2 cm of substrate was collected with a flat shovel from each pond. The substrate was collected from random locations in the pond until 0.01 m^3 of substrate was collected. The samples were homogenized by adding water and gently stirring the mixture. The assumption was made that the substrate samples contained viable clam shrimp eggs. The homogenized substrate was placed in 28 $\times$ 35-cm pans. Sediment samples were subjected to five different drying and freezing treatments:

Treatment A – The samples were not dried.
Treatment B – The samples were air dried in the sun for 2 days.
Treatment C – The samples were air dried in the sun for 4 days.
Treatment D – The samples were air dried in the sun for 21 days.
Treatment E – The samples were air dried in the sun for 4 days and frozen for 7 days.

Five randomly selected, 100-cc sediment samples were removed from the drying pans once the treatments were completed, placed in 600-mL beakers, and covered with 500 mL of well water. Clam shrimp that emerged were enumerated. Analysis of variance followed by a least squares means procedure (SAS Institute, Inc. 1988) was used to determine if differences in numbers of clam shrimp hatched in each treatment were statistically significant ($P = 0.05$).

RESULTS AND DISCUSSION

Hatchery Ponds

Clam shrimp densities were greatest in ponds subjected to a drying period prior to being filled in the spring (Treatments 1 and 2) (Table 1). Treatment 2 ponds, which were drained and dried in the spring, had significantly ($P = 0.001$) more clam shrimp than the other treatments. These ponds had such dense numbers of clam shrimp that treatment with Batex was necessary during the production season. Treatment 1 ponds, which were left empty during the winter and were filled just prior to stocking, had significantly ($P = 0.001$) more clam shrimp than Treatment 3 and 4 ponds.

Treatment 3 and 4 ponds, which were not dried in spring, had few clam shrimp (Table 1). These results are similar to Dexter and McCarraher's (1967) results. They reported that winter drying and the early spring culture of northern pike, *Esox lucius*, promoted hatching, growth, and reproduction of clam shrimp.

Turbidity induced by Eubranchiopoda is a frequently reported problem (Hornbeck et al. 1965; McCraren and Phillips 1977; McCraren et al. 1977). In this study, high turbidities did occur in ponds with high numbers of clam shrimp; however, high turbidities

TABLE 1. Mean number/m^2 of clam shrimp, mean turbidity in Jackson Turbidity Units, and mean number of walleye fingerlings harvested and percent survival in 0.4-ha earthen ponds at Blind Pony Hatchery, Missouri for each of four different water management treatments (see text for explanation of treatments). Mean values of clam shrimp followed by different letters are significantly different ($P = 0.001$).

	Treatment 1	Treatment 2	Treatment 3	Treatment 4
No. of clam shrimp	470b	12,835a	3.5c	0.5c
Turbidity	14	53	45	19
Walleye No. harvested	4,900	1,500	78,650	115,150
Percent survival	1.7	0.5	26.5	38.5

also occurred in ponds with few or no clam shrimp. Treatment 2 ponds, which had the highest number of clam shrimp, had the highest turbidities. High turbidities also occurred in Treatment 3 ponds, which had very few clam shrimp. The cause of turbidity in Treatment 3 ponds was believed to be burrowing mayflies, *Hexagenia* sp., that were present in high numbers.

Hornbeck et al. (1965) and Dexter and McCarraher (1967) reported that large numbers of Eubranchiopoda can reduce fish production in hatchery ponds. This problem has also been observed in Missouri Department of Conservation hatcheries and was found to be true during this experiment.

The ponds with the highest number of clam shrimp produced the lowest numbers of walleye fingerlings (Table 1). Production in Treatment 1 ponds (dry over winter) and Treatment 2 ponds (dry in spring) averaged 4,900 and 1,500 fish/ha, with only 1.7% to 0.5% survival, respectively. Treatment 3 ponds, which were dry after harvest but full in winter and spring, produced an average of 78,650 fish/ha, (26.5% survival). One of the Treatment 4 ponds, which were not dried, produced more walleye than any other pond (226,000/ha), but the other produced only 4,300 fish/ha for an average of 115,150 fish/ha. The difference in production between these two ponds is puzzling and cannot be explained.

LABORATORY

Weaver (1943), Prophet (1963), Bishop (1967), Dexter and McCarraher (1967), Grainger (1991), and Hann and Lonsberry (1991) reported drying and freezing stimulated or enhanced the hatching of Eubranchiopoda eggs. In the laboratory study, drying and freezing increased the number of clam shrimp that hatched (Table 2). Significantly ($P = 0.05$) more clam shrimp hatched in Treatment E of Pond 1 than in other Pond 1 treatments. In Pond 2, significantly ($P = 0.01$) more clam shrimp hatched in Treatment E than from other Pond 2 treatments, except for Treatment C.

Hatchability information obtained from substrate dried in pans is not directly applicable to hatchery ponds. Drying in a pan was controlled, and no moisture was added once drying started. In a pond, moisture can come from numerous sources. Saturated soil

TABLE 2. Mean number of clam shrimp hatched per 600-mL beaker from substrate that was dried and frozen. Mean values in a column followed by different letters are significantly different ($P = 0.05$)

Treatment	Pond 1	Pond 2
A (wet; not dried)	0b	0b
B (dried for 2 days)	0.4b	1.0b
C (dried for 4 days)	0b	9.0a
D (dried for 21 days)	1.7b	1.6b
E (dried and frozen)	5.0a	8.0a

around a pond allows water to seep into a pond for several days, and dew and precipitation can also add moisture.

The results of this study suggest that elimination of freezing and drying during the winter and/or early spring should reduce or eliminate clam shrimp from hatchery ponds. In ponds which have heavy infestations of clam shrimp, it is recommended that the pond be dried immediately after draining and harvest and then refilled in fall or early winter and left full until spring.

Filling ponds in the fall or early winter may make it difficult to establish a good zooplankton population the following spring because drying over winter promotes hatching of zooplankton eggs, just as it promotes hatching of clam shrimp eggs. Draining and immediately refilling ponds just before fry are stocked may be helpful in establishing zooplankton populations. In the absence of winter drying, insect, vegetation, or disease problems may develop. If this happens, ponds can be dried during winter and early spring every second or third year to deal with these problems. Clam shrimp may reappear; however, numbers should be low.

Using water management to control clam shrimp offers hatchery managers another option in dealing with a persistent and serious problem. However, managers need to evaluate the benefits and the problems of not allowing ponds to dry over winter. Clam shrimp and other Eubranchiopoda can be reduced, but this may cause other problems which interfere with fish production.

ACKNOWLEDGMENT

Funds for this study were provided in part by the Federal Aid in Fish and Wildlife Restoration Act, Missouri Project F-1-R.

REFERENCES

Bishop, J.A. 1967. Some adaptations of *Limnadia stanleyana* King (Crustacea: Branchiopoda: Conchostraca) to a temporary freshwater environment. Journal of Animal Ecology 36:599-609.

Conover, W.J. 1980. Practical Nonparametric Statistics, 2nd ed. John Wiley and Sons, Inc., New York, New York.

Dexter, R. W., and D. B. McCarraher. 1967. Clam shrimps as pests in fish rearing ponds. Progressive Fish-Culturist 29:105-107.

Grainger, J.N.R. 1991. The biology of *Tanymastix stagnalis* (L.) and its survival in large and small temporary water bodies in Ireland. Hydrobiologia 212:77-82.

Hann, B.J., and B. Lonsberry. 1991. Influence of temperature on hatching of eggs of *Lepidurus couesii* (Crustacea, Notostraca). Hydrobiologia 212:61-66.

Hornbeck, R.G., W. White, and F.P. Meyer. 1965. Control of *Apus* and fairy shrimp in hatchery rearing ponds. Proceedings of the Southeastern Association of Game and Fish Commissioners 19:401-403.

McCraren, J. P., and T. R. Phillips. 1977. Effects of Masoten (Dylox) on plankton in earthen ponds. Proceedings of the Southeastern Association of Fish and Wildlife Agencies 31:441-448.

McCraren, J. P., J. L. Millard, and A. M. Woolven. 1977. Masoten (Dylox) as a control for clam shrimp in hatchery production ponds. Proceedings of the Southeastern Association of Fish and Wildlife Agencies 31:329-331.

Pennak, R. W. 1953. Freshwater Invertebrates of the United States. The Ronald Press Co., New York, New York.

Prophet, C. 1963. Some factors influencing the hatching of anostracan eggs. Transactions of the Kansas Academy of Science 62:153-161.

SAS Institute, Inc. 1988. SAS Language Guide for Personal Computers, Release 6.03 ed. SAS Institute, Inc., Cary, North Carolina.

Weaver, C. R. 1943. Observations on the life cycle of *Eubranchipus vernalis*. Ecology 24:500-502.

Whiteside, M. C. 1974. Chydorid (Cladocera) ecology: Seasonal patterns and abundance of populations in Elk Lake, Minnesota. Ecology 55:538-550.

Impact of Predation by Backswimmers in Golden Shiner, *Notemigonus crysoleucas,* Production Ponds

Joseph G. Burleigh
Robert W. Katayama
Nader Elkassabany

ABSTRACT. Laboratory and field studies were conducted to determine the population dynamics and potential for predation on golden shiner, *Notemigonus crysoleucas,* fry by the backswimmers, *Notonecta indica* and *Buenoa scimitra.* Both species were found to be highly effective predators under laboratory conditions. *N. indica* invaded newly flooded ponds within 2 weeks of putting spawning mats with eggs into the ponds and were thus present when fry were vulnerable to predation. *B. scimitra* invaded ponds at a slower pace and would not be as likely to cause extensive predation. *N. indica* had significantly lower ($P = 0.05$) population levels in ponds that had been established for one year. The potential loss from backswimmer predation could easily exceed 2.3 kg per 100 m of shoreline, assuming 40% of the fry lost to predation would survive to harvest. Diesel

Joseph G. Burleigh, Robert W. Katayama, and Nader Elkassabany, Department of Agriculture, University of Arkansas at Pine Bluff, Pine Bluff, AR 71601, USA.

Correspondence for Nader Elkassabany may be addressed to Department of Entomology, University of Arkansas, Fayetteville, AR 72701, USA.

[Haworth co-indexing entry note]: "Impact of Predation by Backswimmers in Golden Shiner, *Notemigonus crysoleucas,* Production Ponds." Burleigh, Joseph G., Robert W. Katayama, and Nader Elkassabany. Co-published simultaneously in the *Journal of Applied Aquaculture,* (The Haworth Press, Inc.) Vol. 2, No. 3/4, 1993, pp. 243-256; and: *Strategies and Tactics for Management of Fertilized Hatchery Ponds* (ed: Richard O. Anderson and Douglas Tave) The Haworth Press, Inc., 1993, pp. 243-256. Multiple copies of this article/chapter may be purchased from The Haworth Document Delivery Center [1-800-3-HAWORTH; 9:00 a.m. - 5:00 p.m. (EST)].

243

fuel treatment of backswimmer populations in 0.05-ha ponds resulted in about a 50% control of *N. indica*, mostly during the first 24 hours. *B. scimitra* were not as susceptible to the diesel fuel treatment. Several classes of insecticides were evaluated in the laboratory as potential control agents; Curacron, an organophosphate, and Cymbush a, synthetic pyrethroid, produced 100% mortality of *N. indica* at a concentration of 0.01 ppm. *B. scimitra* was even more sensitive to these compounds.

INTRODUCTION

Commercial production of baitfish accounts for about 35% of total aquaculture sales in Arkansas. In 1989, 3,245,000 kg of golden shiner, *Notemigonus crysoleucas,* valued at $19,635,000 were produced on 8,256 ha, and an additional 2,752 ha produced $6,463,000 worth of other baitfish (Anon. 1991).

Baitfish production averages 393 kg/ha (Anon. 1991) but ranges up to 1,348 kg/ha (Engle et al. 1990). Factors contributing to low production include poor water quality, pathogens, and predators. Production practices in the baitfish industry are mostly based upon techniques developed by the growers. Traditionally, producers stock their ponds by egg transfer (Giudice et al. 1982; Dorman and Gray 1987). Fiber mats laden with golden shiner eggs are placed into clear, shallow basins in partially flooded ponds, where water depth varies from 2 to 20 cm. These ponds are initially characterized by a smooth mud bottom with little or no vegetation and few physical structures. Eggs hatch in 2-3 days, after which the fry attach themselves to the mats for a day or so prior to reaching the free-swimming stage. Free-swimming fry frequent the pond's edge for 2-3 weeks before moving into deeper water. Throughout this developmental period, ponds are continually filled with well water or water from an adjacent pond; thus, the depth and margin constantly fluctuate until ponds are full in about 3 weeks. The draining and refilling sequence was designed to eliminate the presence of unwanted aquatic plants and animals, particularly vertebrate predators. However, several species of insects are found in the vicinity of small fry. Many producers believe that these insects prey on the fry (Berezina 1956; Barr et al. 1978; Cronin and Travis 1986) and treat their ponds to control these insects.

Backswimmers are insects that may be significant predators. A series of studies was initiated to determine the role that backswimmers play in commercial minnow ponds and to understand the biology of these insects (Giller and McNeil 1981). Several important factors must be determined to understand the role of backswimmers in these ponds. First, it must be determined whether the insects are present in the ponds at the appropriate time and in sufficient numbers to cause problems (Bardach et al. 1972). Secondly, it must be determined which backswimmer species prey upon fry, and their predation rates must be determined.

Traditional insect control procedures in baitfish ponds have evolved without real scrutiny of their impacts upon insect populations. The use of diesel fuel or motor oil, the currently employed procedure (Giudice et al. 1982), is the traditional treatment, and it may be both environmentally polluting and inefficient. Thus, laboratory and field evaluations of diesel fuel were examined in order to determine its effectiveness. The use of insecticides in aquatic environments is rather restricted (Schnick et al. 1986), but the insecticides might have potential, because these fish are not grown for human consumption. Therefore, laboratory evaluation of several insecticides was also undertaken.

MATERIALS AND METHODS

Field Studies

Populations of backswimmers, *Notonecta indica* and *Bueona scimitra*, were monitored in four commercial minnow ponds in Arkansas. The ponds ranged in area from 6.1 to 8.1 ha. Two ponds were started filling in mid-May, 1988. Mats with eggs were placed in the ponds 5 days after filling started. Two other ponds were started filling in mid-May, 1989 and were stocked with eggs as in 1988. Sampling commenced when golden shiner fry were observed near the mats (week 1), i.e., about 10 days after filling started in ponds. Ponds were sampled each week until the end of September, 1989.

Sampling for *N. indica* consisted of observing a 2 × 2 m area of pond surface for 5 minutes at sites on the north, south, east, and west banks of each pond. Numbers counted at each area were combined and included both adult and immature forms. *B. scimitra* nymphs and adults were sampled with a triangular (30.5 cm) aquatic insect net. The net was extended about 1.5 m from shore and then retracted. Eight net pulls were made at each of two locations in a pond.

To monitor the effect of diesel fuel on population levels of back-swimmers, two 0.05-ha earthen ponds were drained in late-June and again in early August and were then flooded but were not stocked with fish. After about 3 weeks, diesel fuel was applied at the recommended rate of 4.6 L/ha on the upwind side of the pond at a wind speed of approximately 224 cm/second. Diesel fuel treatments were evaluated by sampling the population prior to treatment and 1, 24, 48, and 72 hours after treatment in both replications. Sampling consisted of eight aquatic net pulls on opposite sides of each pond.

Laboratory Studies

Laboratory tests were conducted at room temperature (22°C) with 16 hours of light and 8 hours of dark, to determine the predatory potential of nymphal and adult *N. indica* and adult *B. scimitra.* The tests were conducted in 3.8-L plastic vessels; they were filled to a depth of 12 cm (2L) with charcoal-filtered water. This water depth was chosen because it was representative of the pond depth approximately 1m from the bank. No structure was provided to enable the fry or insects to hide, because generally no structure was available in the ponds. Ten 5- to 8-mm golden shiner fry were placed in each jar along with one insect. There were six different treatments for *N. indica,* one for each of its six life stages, and one treatment for the adult stage of *B. scimitra.* Each treatment was replicated five times for *N. indica* and three times for *B. scimitra.* A control treatment was conducted for each life stage tested. Dead fry were recorded after 24 hours for *N. indica* and after 24-, 48- and 72-hour periods for *B. scimitra.*

To obtain known nymphal stages, eggs from adult *N. indica* were collected from plastic-coated wire placed in aquaria containing adult backswimmers. These eggs were placed in separate aquaria where they were hatched and raised. Following hatching or molting, the appropriate life stages were placed in the 3.8-L plastic jars. Golden shiner fry were obtained from the fish hatchery of the Aquaculture Research Station at the University of Arkansas at Pine Bluff (UAPB). Fry varied in age from D5 to D15.

Diesel Fuel Experiments

These tests were conducted at room temperature (22°C), with 16 hours of light and 8 hours of dark. Four 3.8-L glass vessels were filled with 1L of charcoal filtered water. Adult *N. indica* and *B. scimitra* were collected from a pond at UAPB for these experiments. A single adult backswimmer was placed in a 3.8-L vessel (177 cm^2). Several hours later, diesel fuel was applied at 1 ml, 0.1 ml, or 0.01 ml which was equivalent to 1,500, 150 and 15 L/ha, respectively, in each of 10 replicates for both species; control vessels contained filtered water and an adult backswimmer. Backswimmer mortality was recorded at 1, 24, and 48 hours.

Insecticide Screening

Insecticide evaluations were conducted in the same manner as the diesel fuel tests. The insecticide concentrations varied from 1,000 ppm to 0.0001 ppm for each of the following insecticides:[1] Pydrin, Curacron, Cymbush, Scout, Bolstar, Baytex, Asana, Dimilin, and Larvin. These insecticides were chosen because they are commonly used on agricultural pests in southeast Arkansas. Field grade insecticides were employed rather than the technical grade. *N. indica* and *B. scimitra* adults were collected from ponds at the UAPB experiment station, transported to the laboratory, and placed in the insecticide solution. Insecticides were serially diluted 10-fold from 1,000 ppm to 0.0001 ppm. The tests consisted of two replicates for each treatment and control; there were 10 backswimmer adults per treatment for each species. Mortality was recorded at 1-, 24-, and 48-hour intervals.

Statistical Analysis

Differences among treatments were evaluated using analysis of variance and Duncan's multiple range test. The Statistical Analysis System was employed for all analysis (SAS Institute, Inc. 1985).

---

1. Use of trade names does not imply endorsement.

RESULTS AND DISCUSSION

N. indica and *B. scimitra* were observed occupying different strata in the ponds. *N. indica* inhabited the pond surface within 2 m of the shoreline. They either cruised in search of prey or clung to plants or other structures from which they attacked 5- to 30-mm golden shiner fry. *B. scimitra* swam below the surface and extended to at least 5 m from shore. These behavioral differences affected sampling efficacy and the relative population densities based upon the two sampling methods. Therefore, visual sampling was used for *N. indica* and aquatic net sampling used for *B. scimitra*.

Adult *N. indica* rapidly invaded the recently flooded ponds (Figure 1) as was noted by Walton et al. (1990). There were detectable numbers of adults, and eggs were observed on the mats and vegetation within 2 weeks of golden shiner fry hatching; thus, these insects were present when fry were at a vulnerable stage. Sampling counts peaked in the last week of June. This dramatic increase was due to a new cohort resulting from eggs laid by the colonizing adults. In July, there was a rapid decrease in the numbers observed. The data indicate that there was at least one generation of *N. indica* during the summer period. Newly emerged nymphs were seen in small

FIGURE 1. Average number of *Notonecta indica* collected at four locations in study ponds. The established ponds were flooded beginning in mid-May, 1988 and the recently flooded ponds were flooded in mid-May, 1989.

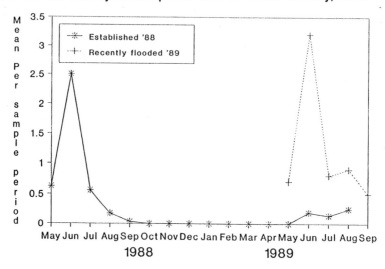

numbers throughout the remainder of summer but did not appear to be the product of a well-defined cohort.

In May, adult *N. indica* were observed hunting in the shallows among young golden shiner fry. They were observed making numerous lunges in the vicinity of the fry; however, the small fry (5 mm) could not be visually detected in the insects' grasp.

Adjacent ponds, which had been flooded for one year and which held mature golden shiners, did not experience as rapid an influx of insects and had significantly lower ($P = 0.05$) population levels throughout the summer (Figure 2). This difference could have been the result of reduced "attractiveness" of the ponds or the result of predation by adult golden shiners on *N. indica*.

B. scimitra appeared in the recently flooded ponds in measurable numbers between the second and third weeks (Figure 3). The population then increased gradually until a new generation of nymphs appeared in late July (week 10). In this study, *B. scimitra* were collected in relatively low numbers during May and June when

FIGURE 2. *Notonecta indica* population levels throughout the summer in established and recently flooded ponds. The established ponds were flooded in mid-May, 1988 and the recently flooded ponds were flooded in mid-May, 1989.

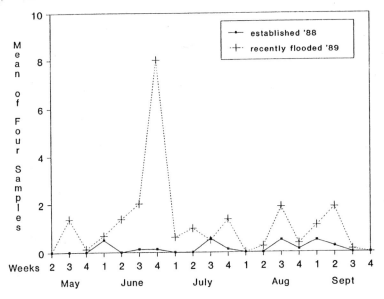

golden shiner fry were probably most vulnerable. Thus, predation would be low; however, *B. scimitra* readily feed upon microcrustaceans and immature aquatic insects (Toth and Chew 1972); therefore, they could be formidable competitors for food during the mid- to late-summer months. There was no significant difference between the number sampled in recently flooded ponds and those which had been flooded for a year (Figure 3), as was the case for *N. indica*. This indicated that established ponds were as attractive and as suitable for reproduction as were the recently flooded ponds and that predation from mature golden shiners was not a factor.

Field studies showed that both backswimmers were present and active in the shallows at the time when fry were present. However, actual predation could not be observed or quantified, because the small fry could not be observed in the clutches of the insects. Therefore, laboratory experiments were used to test the potential for predation. First instar nymphs were the only stage of *N. indica* that did not kill a majority of or all of the fry (Table 1). Under these

FIGURE 3. *Buenoa scimitra* population levels throughout the summer in established and recently flooded ponds. The established ponds were flooded in May, 1988 and the recently flooded ponds were flooded in mid-May, 1989.

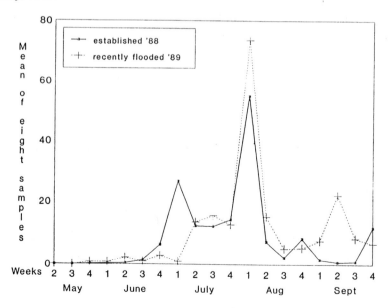

TABLE 1. Golden shiner fry killed in 24 hours by a single *Notonecta indica* nymph or adult when placed in 2 L of water stocked with 10 golden shiner fry. Means followed by the same letter were not significantly different (*P* = 0.05).

		Mortality					
Replication	Control	Instar 1	Instar 2	Instar 3	Instar 4	Instar 5	Adult
1	0	0	5	10	10	10	10
2	0	0	3	10	10	10	10
3	0	2	8	10	10	10	10
4	1	1	8	10	10	10	10
5	0	0	10	10	10	10	10
Mean	0.2a	0.06a	6.8b	10.0c	10.0c	10.0c	10.0c

controlled conditions, *N. indica* proved to be a highly effective predator and was capable of killing at least 10 fry/day. *N. indica* are highly mobile and can seek new prey when the resource in a particular area of a pond is diminished; thus, predation in the field could be considerably greater. Immature nymphal stages of *N. indica* are not capable of inter-pond migration; therefore, most of these insects present in recently flooded ponds would be adults, unless there was a delay beyond the usual 4-5 days in placing the egg-laden mats into the ponds. A delay in matting the newly flooded ponds of 1-2 weeks would lead to the presence of mature and immature *N. indica*, thus increasing the potential for predation.

Adult *B. scimitra* preyed upon golden shiner fry under laboratory conditions (Table 2), but they required a longer period to kill fry. Low population levels in ponds during weeks 1-3 indicated that their contribution to predation was probably minimal. However, the potential for predation by *B. scimitra* is there (Stewart and Miura 1978) should they immigrate more rapidly or in greater numbers during earlier or later periods of the matting season, e.g., April or June.

TABLE 2. Average number of dead golden shiner fry in three replications of 10 fry in 2L of water with one adult *Buenoa scimitra* after 24, 48 and 72 hours. Means in a row followed by the same letter are not significantly different ($P = 0.05$).

Hours	Mortality	
	Control	Average of all replications
24	0.0a	$6.5 \pm 1.25b$
48	1.5a	$9.5 \pm 0.25b$
72	3.5a	$10.0 \pm 0.0b$

The relatively low numbers of adult *N. indica* seen in the ponds during the first and second weeks after the fry hatched, as compared to weeks 3-4, suggest a moderate potential for predation. Based on data collected by aquatic net and visual sampling at approximately 2 weeks post hatch, there were at least 30 and 40 adult *N. indica* per 100 m of shoreline, respectively. If one used an estimate of 10 fry consumed per day (a conservative estimate) and considered that there was a 2-week period in which fry were vulnerable, the number of fry lost per 100 m of shoreline could range from 4,333 to 5,666, respectively. Recent studies (Morrison and Burtle 1989) have shown that from 40 to 70% of fry survive to fingerling size. If 40% of the "lost" fry might be expected to survive, the added production might be 2.3 to 2.7 kg of mature shiners or an added value of $18-20 per 100 m of shoreline.

Since it appeared that predation occurred at an economically important level, an evaluation of control procedures was warranted. Laboratory and field studies were conducted to ascertain the effect of diesel fuel on backswimmers. Tests revealed that even under highly stable conditions diesel fuel had a variable effect, depending upon species and exposure time. The 0.01-ml treatment, which was equivalent to 15 L of diesel fuel/ha, did not give complete control (Table 3), even under 48 hours of continual exposure in a stable environment, a highly unlikely condition in the field. *N. indica*'s

TABLE 3. Percent mortality of *Notonecta indica* and *Buenoa scimitra* from various amounts of diesel fuel applied to 1 L (177 cm^2) of water.

Species	Treatment Amount	Control	Mortality		
			1 hour	24 hour	48 hour
Notonecta indica	1.0 ml	0	100	100	100
	0.1 ml	0	88	100	100
	0.01 ml	0	9	70	80
Buenoa scimitra	1.0 ml	0	100	100	100
	0.1 ml	5	70	70	73
	0.01 ml	5	5	10	10

FIGURE 4. Number of *Notonecta indica* and *Buenoa scimitra* sampled prior to and after treatment with 4.6 L/ha of fuel oil. Number is the average collected with an aquatic net.

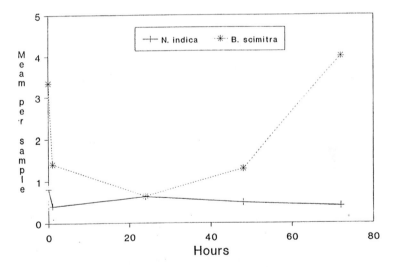

surface-inhabiting behavior made it more vulnerable than *B. scimitra*, which is usually suspended within the water column.

When tested in 0.05-ha earthen ponds, diesel fuel treatments reduced number of *N. indica* sampled by about 50% in the first hour (Figure 4) but had little effect thereafter. There was a similar effect upon *B. scimi-*

TABLE 4. The lowest concentration of an insecticide in ppm that killed 100% of adult *Buenoa scimitra* and *Notonecta indica* in 48 hours.

Insecticide	Buenoa scimitra	Notonecta indica
ORGANOPHOSPHATES		
Curacron	0.0001	0.01
Bolstar	0.1	1.0
Baytex	0.1	10.0
CARBAMATES		
Larvin	–	10.0
CHLORINATED DIPHENYL		
Dimilin	1,000.0	1,000.0
SYNTHETIC PYRETHROIDS		
Pydrin	0.0001	1.0
Cymbush	0.001	0.01
Scout	0.01	0.01
Masoten	1,000.0	1,000.0
Asana	–	0.01

tra; however, the number of *B. scimitra* per sample rebounded in 2 days to levels equivalent to pre-treatment, indicating either a sampling error or recruitment during this period. In any case, diesel treatment had only a moderate impact on backswimmer populations levels.

The poor level of control obtained with diesel encouraged the screening of several insecticides as alternate means of controlling backswimmers. *B. scimitra* appeared to be more susceptible to the insecticides (Table 4); however, Curacron and Cymbush gave 100% control of both species at very low concentrations. Such results warrant further investigation. However, the difficulty with federal registration of these products as a control for backswimmers is formidable.

ACKNOWLEDGMENTS

We gratefully acknowledge the assistance of Neal Anderson, Anderson's Fish Farm, Lonoke, Arkansas for permission to study his golden shiner ponds. The authors also thank Nathan Stone, Owen Porter, and Kwang Lee for their critical review of the manuscript. This article is published with the approval of the Director, Arkansas Agricultural Experiment Station. The research was supported by Cooperative States Research Service Evans/Allen, PL 95-113 grant, AR.X. 1257 to the Department of Agriculture, University of Arkansas at Pine Bluff.

REFERENCES

Anon. 1991. State Plan for Arkansas Aquaculture: Current Status and Potential Development. University of Arkansas at Pine Bluff and University of Arkansas Cooperative Extension Service, Pine Bluff, Arkansas.

Bardach, J. E., J. H. Ryther, and W. O. McLarney. 1972. Aquaculture: The Farming and Husbandry of Aquatic Organisms. John Wiley and Sons, Inc., New York, New York.

Barr, J. F., J. V. Huner, D. P. Klarberg, and J. Witzig. 1978. The large invertebrate-small vertebrate fauna of several small south Louisiana crawfish ponds with emphasis on predaceous arthropods. Proceedings of the World Mariculture Society 9:683-700.

Berezina, N. A. 1956. The use of insecticides to control predaceous insects which are enemies of fish. Voprosy Ikhtiologii 7:209-220. (Translated by the Fisheries Board of Canada).

Cronin, J. T., and J. Travis. 1986. Size-limited predation on larval *Rana areolota* (Anura: Ranidae) by two species of backswimmer (Insecta: Hemiptera: Notonectidae). Herpetologica 42:171-174.

Dorman, L. W., and D. L. Gray. 1987. Spawning Baitfishes. University of Arkansas Cooperative Extension Service, FS 9003, Little Rock, Arkansas.

Engle, C. R., L. W. Dorman, and D. L. Gray. 1990. Baitfish Production, Enterprise Budget. FS 9016, University of Arkansas Cooperative Extension Service, Little Rock, Arkansas.

Giller, P.S., and S. McNeil. 1981. Predation strategies, resource partitioning and habitat selection in *Notonecta* (Hemiptera/Heteroptera). Journal of Animal Ecology 50:789-808.

Giudice, J. J., D. L. Gray, and J. M. Martin. 1982. Manual for Baitfish Culture in the South. Arkansas Cooperative Extension Service Bulletin EC550, Little Rock, Arkansas.

Morrison, J. R., and G. J. Burtle. 1989. Hatching of golden shiner eggs in hatchery tanks and subsequent fry survival in rearing ponds. Progressive Fish-Culturist 51:229-231.

SAS Institute, Inc. 1985. SAS User's Guide-Basic, Version 5th ed. SAS Institute Inc., Cary, North Carolina.

Schnick, R. A., F. P. Meyer, and D. L. Gray. 1986. A Guide to Approved Chemicals in Fish Production and Fishery Resource Management. Arkansas Cooperative Extension Service, Little Rock, Arkansas.

Stewart, R. J., and T. Miura. 1978. Laboratory studies on *Notonecta unifasciata* Guerin and *Buenoa scimitra* Bare as predators of mosquito larvae. California Mosquito Association 46:84-86.

Toth, R. S., and R. M. Chew. 1972. Development and energetics of *Notonecta indica* during predation of *Culex tarsalis* (Diptera:Culicidae). Annals of the Entomological Society of America 65:1271-1279.

Walton, W. E., N. S. Tietze, and M. S. Mulla. 1990. Ecology of *Culex tarsalis* (Diptera:Culicidae): factors influencing larval abundance in mesocosms in Southern California. Journal of Medical Entomology 27:57-67.

Index

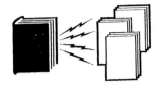

Haworth
DOCUMENT DELIVERY
SERVICE
and Local Photocopying Royalty Payment Form

This new service provides (a) a single-article order form for any article from a Haworth journal and (b) a convenient royalty payment form for local photocopying (not applicable to photocopies intended for resale).

- *Time Saving:* No running around from library to library to find a specific article.
- *Cost Effective:* All costs are kept down to a minimum.
- *Fast Delivery:* Choose from several options, including same-day FAX.
- *No Copyright Hassles:* You will be supplied by the original publisher.
- *Easy Payment:* Choose from several easy payment methods.

Open Accounts Welcome for . . .
- Library Interlibrary Loan Departments
- Library Network/Consortia Wishing to Provide Single-Article Services
- Indexing/Abstracting Services with Single Article Provision Services
- Document Provision Brokers and Freelance Information Service Providers

MAIL or *FAX* THIS ENTIRE ORDER FORM TO:

Attn: **Marianne Arnold**
Haworth Document Delivery Service
The Haworth Press, Inc.
10 Alice Street
Binghamton, NY 13904-1580

or FAX: (607) 722-1424
or CALL: 1-800-3-HAWORTH
(1-800-342-9678; 9am-5pm EST)

PLEASE SEND ME PHOTOCOPIES OF THE FOLLOWING SINGLE ARTICLES:
1) Journal Title: _____
    Vol/Issue/Year:_____Starting & Ending Pages:_____
Article Title:_____
_____

2) Journal Title: _____
    Vol/Issue/Year:_____Starting & Ending Pages:_____
Article Title:_____
_____

3) Journal Title: _____
    Vol/Issue/Year:_____Starting & Ending Pages:_____
Article Title:_____
_____

4) Journal Title: _____
    Vol/Issue/Year:_____Starting & Ending Pages:_____
Article Title:_____
_____

(See other side for Costs and Payment Information)

COSTS: Please figure your cost to order quality copies of an article.

1. Set-up charge per article: $8.00
   ($8.00 × number of separate articles) _____

2. Photocopying charge for each article:
   - 1-10 pages: $1.00 _____
   - 11-19 pages: $3.00 _____
   - 20-29 pages: $5.00 _____
   - 30+ pages: $2.00/10 pages _____

3. Flexicover (optional): $2.00/article _____

4. Postage & Handling: US: $1.00 for the first article/
   - $.50 each additional article _____
   - Federal Express: $25.00 _____
   - Outside US: $2.00 for first article/
   - $.50 each additional article _____

5. Same-day FAX service: $.35 per page _____

6. Local Photocopying Royalty Payment: should you wish to copy the article yourself. Not intended for photocopies made for resale. $1.50 per article per copy (i.e. 10 articles x $1.50 each = $15.00) _____

GRAND TOTAL: _____

METHOD OF PAYMENT: (please check one)

❏ Check enclosed   ❏ Please ship and bill. PO # _____
(sorry we can ship and bill to bookstores only! All others must pre-pay)

❏ Charge to my credit card: ❏ Visa; ❏ MasterCard; ❏ American Express;

Account Number:_____ Expiration date:_____

Signature: X_____ Name: _____

Institution: _____ Address: _____

City: _____ State:_____ Zip:_____

Phone Number: _____ FAX Number: _____

MAIL or *FAX* THIS ENTIRE ORDER FORM TO:

Attn: **Marianne Arnold**
Haworth Document Delivery Service
The Haworth Press, Inc.
10 Alice Street
Binghamton, NY 13904-1580

or FAX: (607) 722-1424
or CALL: 1-800-3-HAWORTH
(1-800-342-9678; 9am-5pm EST)